John Woodbridge Davis

Formulae for the Calculation of Railroad Excavation and

Embankment

John Woodbridge Davis

Formulae for the Calculation of Railroad Excavation and Embankment

ISBN/EAN: 9783744727075

Printed in Europe, USA, Canada, Australia, Japan

Cover: Foto ©berggeist007 / pixelio.de

More available books at **www.hansebooks.com**

FORMULÆ

FOR THE CALCULATION OF

RAILROAD

EXCAVATION AND EMBANKMENT

BY

JOHN WOODBRIDGE DAVIS,

CIVIL ENGINEER.

USED AS A TEXT-BOOK IN THE
SCHOOL OF MINES,
COLUMBIA COLLEGE.

———◆◆———

NEW YORK:
J. DICKSON & BRO., BOOK AND JOB PRINTERS,
13 NORTH WILLIAM STREET.
1876.

PREFACE.

THE METHOD of calculating earthwork discussed in this volume originated with the necessities for expedition and accuracy, while the author was engaged in making extensive computations of this kind, and from actual study of the ground and the manner in which irregularities extend from one cross-section to another, made with the express purpose of obtaining general and accurate formulæ for all shapes that occur in the engineer's practice. Thus the several portions forming the entire plan, which the author now considers complete, were gradually developed in about the order in which they appear.

Although it was necessary to consult many works on mensuration and civil-engineering in order to be assured of the novelty of some of the devices, and to curtail the description of others commonly known, direct use has been made of two only. The formula of Henck for correcting the calculation of earthwork on curves has been adopted; and in the note at the end of this treatise the hint has been taken from Professor Gillespie of applying integral calculus to railroad volumes, though this would seem a natural use of that department of mathematics. In both instances, the source and value of the aid is thankfully acknowledged.

In the formula for the area of the mere ordinary cross-section there could not be much originality. The rate of slope has been included, and the values that vary from section to section reduced to two. Also the plan of combining consecutive volumes of equal length in series is one of the first that present themselves to calculators. Beyond this each topic is presented as new. These are, first, the inclusion of volumes of minor length in the secutive volumes forms no interruption to a single and speedy operation. Next, the attainment of a single for mula for all varieties of irregular and defective volumes which not only abbreviates the computation of each, bu allows them all to be included in the same series witl ordinary volumes, so that any entire cut or bank can be estimated accurately in one operation. Approximate formulæ, on account of their simplicity, are used in the combinations, and the results corrected by an exceed ingly easy formula, thus obtaining the contents as by the prismoidal rule. Formulæ of error are often met witl in discussions on mensuration. Prof. Gillespie has in vestigated formulæ for the errors of all common approx imate methods, using the prismoid 1 ormula as a stand ard; but the shape he treats is not very frequent in earth work, and he does not propose to use the formulæ as corrections. In SONNET'S DICTIONNAIRE DES MATHÉMA TIQUES APPLIQUÉES occurs a formula of difference for earth work volumes, identical with that used in this treatise but it is exhibited as a formula of error, not of correc tion. For the similar formula to be applied to irregula volumes the author has seen no counterpart. The mode of calculating level-section volumes differs slightly from the best methods in use, one column in the opera tion being constructed by a simpler rule than the ordi nary. Complete formulæ for the finished work, exactly similar to the formulæ for the original computation, and capable of being joined in series, without interruption through a whole cutting, whatever its irregularity, i believed to be entirely new. Ordinarily, the origina calculations are disregarded as useless for the fine estimation, unless the work has been finished exactly to the prescribed lines, and the advantage of exact rules i lost to the most important calculation of all. A metho is here devised for using the former calculation in con junction with the final, thus removing a large bulk c work from the latter, and giving to it the same advan tage of accuracy as to the other. An easy correctin formula is given for the rude method of assuming th estimates, when the work is not yet staked out. The result is the same as that obtained by approximating with end areas of volumes as actually staked out. Also a method for borrow-pits is described, founded on the principles used for combining road-bed volumes in series.

Although the original idea of the author was to secure the maximum amount of brevity by combining everything without exception in series, and then eliminating every constant factor and term from the calculation of each, inventing for this purpose several devices not hitherto used, yet he has not allowed himself at any point to be tempted away from exact work, deeming this to be of the first importance. The claim for the method is *absolute accuracy* joined with what brevity is shown to arise from the use of the resulting formulæ. Special attention is called to the treatment of irregular volumes, the discussion of which is so provokingly limited and obscure in most books. These are as methodically and exactly calculated as the more regular volumes. A general discussion of this subject will be found in the note at the end of this treatise, where the errors of neglect and false assumption are shown and estimated.

In inviting the attention of engineers to this method, the author considers it fair to state in its behalf the fact that upon carrying, as a perfect stranger, the manuscript of the former part of this work to Gen. Francis L. Vinton, head of the Department of Civil and Mining Engineering, School of Mines, Columbia College, that official immediately adopted it, using it in his lectures of the winter of '75-6, and recommending it, in the shape of a text-book, for use in that Institution. To Gen. Vinton the author must acknowledge his great obligations, for valuable advice in the construction of the work and the manner of presenting the discussions, for important research through our own and foreign methods of computing earthwork, and for that more material aid, without which this volume could not at present have been published by the efforts alone of its

AUTHOR.

FORMULÆ

For the calculation of Railroad Excavation
and Embankment.

TREATMENT OF INTERMEDIATE
. STATIONS.

THE manner of calculating regular cross-sections of
excavation and embankment, contained by uniform
slopes, has been reduced to formulæ by many authors,
representing the operation in concise form ; and these
formulæ, modified by the third dimension, length, have
been moulded to express the content of a volume
between two such sections, and even the bulk of a series,
indefinitely extended, of such volumes, lying consecutive
between cross-sections equi-distant, the width of road-
way and rate of slope, of course, remaining the same
throughout the whole length. We now propose to un-
fold a method of computing by formulæ the contents of
a series uninterrupted by the presence of vols. however
unequal in length, and show the advantage attending
this plan, after reviewing as briefly as possible the
method now in use.

The accompanying diagram represents the cross-
section of a railroad cut, b being half the width of

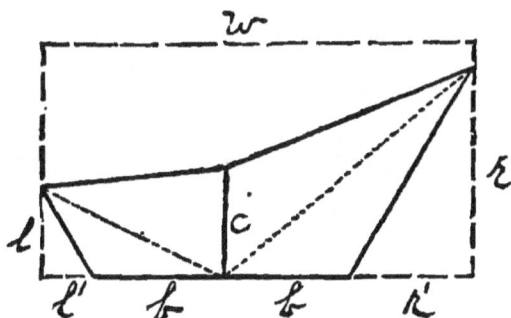

road-bed, c centre-height, r elevation of right slope-stake above grade, r' its horizontal distance from nearest side of road-bed, l elev. left slope-stake, l' its horizontal dist. from nearest side of road-bed, and w the entire top-width or horizontal dist. between slope-stakes. The area of this section is evidently

$$\tfrac{1}{2}rb + \tfrac{1}{2}lb + \tfrac{1}{2}c(r'+b) + \tfrac{1}{2}c(l'+b).$$

Let S denote the ratio of slope: then $S = \dfrac{r}{r'} = \dfrac{l}{l'}$, and $r = Sr'$, $l = Sl'$. Substituting and reducing,

$$Area\ Section = \tfrac{1}{2}Sb(r'+l') + \tfrac{1}{2}c(r'+l'+2b).$$

Adding and subtracting Sb^2 do not change its value;

$$\therefore Area\ Section = \tfrac{1}{2}Sb(r'+l'+2b) - Sb^2 + \tfrac{1}{2}c(r'+l'+2b),$$

or ☞ $$Area\ Section = \tfrac{1}{2}w(c+Sb) - Sb^2.$$

Supposing w', c' to represent width and centre at next station, the area of its cross-section may be expressed by a formula similar to the above: half the sum of these, multiplied by dist., D, between, and divided by 27, gives a near approximate of the volume bet. in cu. yds.

☞ $$Vol. = (wc + w'c' + Sb(w+w') - 4Sb^2)\tfrac{D}{108}.$$

Add two consecutive volumes of equal length by means of the general formula, w'', c'', representing the width and centre of third cross-section:

$$Vol. = (wc + 2w'c' + w''c'' + Sb(w + 2w' + w'') - 8Sb^2)\tfrac{D}{108}.$$

By continual addition we may get a formula for the sum of any number of consecutive volumes; but, letting n denote the number of volumes, we may at once indite a general formula for the calculation of any number of volumes consecutive. Thus we have

$$Vol. = \left\{ \begin{array}{l} wc + 2w'c' + \&c. + 2w_n c_n + w_{n+1}c_{n+1} \\ + Sb(w + 2w' + \&c. + 2w_n + w_{n+1}) - 4Sb^2 n \end{array} \right\} \frac{D}{108}.$$

Divide and multiply by 2 to convert formula into more convenient shape, which now may be expressed

☞ $$Vol. = \left\{ \begin{array}{l} mid\text{-}prods. + \tfrac{1}{2}\ end\text{-}prods. \\ + Sb(mid\text{-}widths + \tfrac{1}{2}\ end\text{-}widths) \\ - 2Sb^2 \times no.\ of\ rods \end{array} \right\} \frac{D}{54}.$$

Let us illustrate this formula by applying it to the following extract from a field book, containing columns of stas., centre cuts, left and right heights and dists. of slope-stakes, the road-bed being 18 ft., slope, 1 to 1, dist. apart of stas., 100 ft.:

STA.	CENTRE.	LEFT.	RIGHT.
1	3.0	$\frac{4.4}{10}$	$\frac{4.0}{13}$
2	5.1	$\frac{5.6}{12}$	$\frac{6.2}{13}$
3	6.4	$\frac{6.0}{13}$	$\frac{7.4}{16}$
4	7.2	$\frac{9.4}{17}$	$\frac{8.0}{17}$
5	9.0	$\frac{9.4}{19}$	$\frac{9.5}{18}$
6	6.7	$\frac{7.3}{16}$	$\frac{6.4}{13}$

OPERATION.

STA.	WIDTHS.		CENTRE.		PRODS.
1	11.55	×	3.0	=	34.65
2	27.8	×	5.1	=	141.78
3	31.5	×	6.4	=	201.60
4	34.1	×	7.2	=	245.52
5	38.0	×	9.0	=	342.00
6	15.85	×	6.7	=	106.195

9 or $Sb \times 158.80$

$$
\begin{array}{l}
1071.745 \\
1429.2 \\
-810. \\
\hline
9)169094.5 \\
6)18788.3 \\
\hline
3131.38 \text{ cu. yds.}
\end{array}
$$

As stated in introductory paragraph, the method of combining consecutive vols. of equal length in series is common; but the formulæ used for this purpose are various. We have been able to find none more concise and fit for combination that those of this paper. For instance, as an example from a popular book, Mr. Henck, to calculate above series, would use a table of ten columns. [See example on page 105 of his FIELD BOOK.] Three of these are cols. of prods.; five cols. must be summed, the first one three separate times, omitting certain values in each addition, the second also three times, and the third twice, making in effect ten cols. to be added. The work under the cols. is

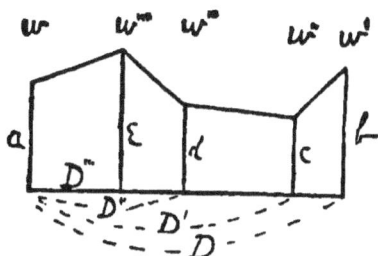

With third interm., e, we have

$$\tfrac{1}{2}D(b+c)+\tfrac{1}{2}D'(d-b)+\tfrac{1}{2}D''(e-c)+\tfrac{1}{2}D'''(a-d).$$

It is noticed that the first terms of these formulæ are identical, being in every case the prod. of half the entire length of vol. by sum of areas of last interm. and last end. The remaining terms we call the corrections for interms., each term being the correction for one interm., having for its coefficient the dist. of that interm. from first end, and for a quantity within the parenthesis the difference of sectional areas next one on each side. This arrangement of values remains unaltered, considering any no. whatever of interms.

Substituting in place of symbols a, b, c, &c., in the expression for three interms. the formula for areas of railroad cross-sections, the positions of w, w', &c., marked in diagram, the centres of same sections accompanied by corresponding strokes, and reducing, we obtain for content of vol. with three interms this expression,

$$\left\{ \begin{array}{l} \left(w''c''+w'c'+Sb(w''+w')-4Sb^2\right)\dfrac{D}{108} \\[2mm] +\left(w'''c'''-w'c'+Sb(w'''-w')\right)\dfrac{D'}{108} \\[2mm] +\left(w''''c''''-w''c''+Sb(w''''-w'')\right)\dfrac{D''}{108} \\[2mm] +\left(wc-w'''c'''+Sb(w-w''')\right)\dfrac{D'''}{108} \end{array} \right\}$$

The expression representing ordinary process is

$$\left\{\begin{array}{l} \left(wc+w''''c''''+Sb(w+w'''')-4Sb^2\right)\dfrac{D'''}{108} \\[2mm] +\left(w''''c'''+w'''c'''+Sb(w''''+w''')-4Sb^2\right)\dfrac{D''-D'''}{108} \\[2mm] +\left(w'''c'''+w''c'''+Sb(w'''+w'')-4Sb^2\right)\dfrac{D'-D''}{108} \\[2mm] +\left(w''c''+w'c'+Sb(w''+w')-4Sb^2\right)\dfrac{D-D'}{108} \end{array}\right\}$$

These expressions contain each four principal terms: of the latter, each term includes the subtraction, $-4sb^2$, while only one term of the former contains this subtraction; and this term, having for coefficient the length of a full vol., D, we intend to combine in series with the other full vols. of cut or fill, where this subtraction will be eliminated from the term and supplied by the common subtraction of the series. Therefore, by this new method of calculating minor vols., the subtraction, $-4sb^2$, otherwise unavoidably present in the formula of each, is virtually eliminated from all. Again, three terms of the former contain the differences of sectional prods. and widths, instead of the sums, a very important feature of brevity in favor of that expression. Furthermore, for coefficients the first expression has D, the length of full vol., always 100 ft. or some other easy factor, and D', D'', D''', the actual dists. of interms. as recorded in field-book; whereas the latter has only one coefficient as noted, the other three being found by subtraction. These points of brevity belong to this method when treating a vol. with single interm., but increase in value with the number of interms.

As lately remarked, the first term, whose coefficient is D, we shall include with the general series, leaving the remaining terms to serve as a correction. This term, as before shown, is the same whatever be the number of interms., always being similar to the formula for the whole vol. except that the prod. and width of last interm. are used instead of those of first end. The formulæ of correction may be reduced to the following

RULE OF COR. for any no. of interms., being the amt. in cu. yds. to be added to the content found by areas of last interm. and last end, multiplied by half the length· of entire vol., to find the true content of a vol.· which has any no. of intermediate cross-sections.

Multiply the dist. of each interm. from first end by the sum of these two differences, viz: the difference of the prod. of the section next before and the prod. of the section next following, and the difference of the sectional width next before and width of the next section following, the latter difference multiplied by Sb.   Divide all by 108.

Now, it is clear that to calculate the contents of a cut or fill we may regard it as having full stas. only, and use the original rule for solution, with the exception that vols. containing interms. must be calculated by means of last interm. and last end cross-sections instead of by end areas; and afterward apply the rule of cor. for interms. just enunciated. It only remains, then, to reconstruct the original rule so as to provide for the above-mentioned exception to the former plan, when we shall have a guide to a very simple and uniform method of calculating the contents of a cut or fill.

If the first vol. of cut or fill contain interms., by the foregoing exception we disregard the width and prod. of first cross-section and use instead those of last interm.: but, if interms. occur in a middle vol., half whose first end-section belongs to the vol. next before, we omit half only of its prod. and width, and use instead those of last interm.   This being the only alteration in the original formula for series of equal-lengthed vols., we shall now restate it, including the exception for interm. sections.

RULE for a near approx. calculation of the entire contents of a cut or fill bet. end full stas.

Add all the midwidths, except those followed by interms., of which add the half-widths increased by the half-widths of last interms.   Add to this sum the half-width of last end-section; and the half-width of first end-section, except it be followed by interms., in which case omit it and add instead the half-width of last interm.   Multiply this sum by Sb.

Add to the above the full prods. of sections where full widths have been used, half-prods. where half-widths have been used, and no prods. where no widths have been used.

From this entire sum subtract 2 Sb² multiplied by the no. of vols. bet. full stas. Multiply remainder by dist. bet. consecutive full stas., and divide by 54.

Apply the rule of · cor. for interms.

Let us apply this rule to the following extract from field-book, where the full stas. are 100 ft. apart, the slope 1 to 1, and roadbed 18 ft. wide.

Sta.	Centre.	Left.	Right.
1	4.1		
+40	5.2		
2	5.4		
3	5.9		
4	6.3		
+27	4.5		
+74	7.2		
+88	7.6		
5	7.0		
6	8.0		
7	8.6		
+19	7.7		
+50	7.4		
8	6.9		
+25	6.8		
+75	6.0		
9	5.1		
10	4.8		

STA.	WIDTHS.		CEN.	PRODS.		CORRECTION.
1		24.8	4.1		101.68	
+40	13.75		5.2	71.5		− 2902
2	27.6	27.6	5.4	149.04	149.04	
3	28.7		5.9	169.33		
4	15.0	30.0	6.3	94.5	189.0	
+27		27.2	4.5		122.4	− 1516
+74		31.8	7.2		228.96	−12996
+88	16.35	32.7	7.6	124.26	248.52	+ 700
5	31.7	31.7	7.0	221.9	221.9	
6	34.0		8.0	272.0		
7	17.7	35.4	8.6	152.22	304.44	
+19		33.2	7.7		255.64	+ 2209
+50	15.45	30.9	7.4	114.33	228.66	+ 2521
8	15.85	31.7	6.9	109.365	218.73	
+25		31.9	6.8		216.92	+ 2363
+75	13.65	27.3	6.0	81.9	163.8	+11358
9	25.1	25.1	5.1	128.01	128.01	
10	11.55		4.8	55.44		
	266.4			1743.795		−17414
				2397.6		+19051
				−1458.		+ 1637

$$+ \quad 8.185$$
$$9)\overline{269158.0}$$
$$6)\overline{29906.4}$$
$$4984.4 \text{ cu. yds,}$$

MODE OF OPERATION.

Make col. of stas. and interms. as in field-book.
Opposite under word WIDTHS make two cols., the second
of full widths to be used for cor., the first of widths and
half widths according to rule.   The rule of cor. for
interms. considers full widths only ; therefore, since the
first col. may contain either full or half widths, it is
convenient to fill a second col. solely with full widths,
where required, especially for the cor., the places where
these full widths are necessary being the sections next
one on each side of each interm. The first col. has,
according to rule, entire widths at mid stas. and half-
widths at end stas., with the exception that, if the first
sta. be followed by interms., the half width of last
interm. is used instead, and, if a mid sta. be followed by
interms., half only of its width is used, and in place of
the other half is used the half width of last interm.
section.   These cols. are formed together by adding
mentally from field-book, setting down the widths and
half widths in each col. as required.   But where so
many interms. occur as in this example it is more
convenient, perhaps, to fill the second col. completely,
since it must be nearly filled as it is, and afterward
transfer the widths and half widths to first col.   Next
under word PRODS. construct two cols., each containing
the prods. of centres by the values in corresponding col.
of widths.   Both these cols. are also formed at once by
multiplying each centre by most convenient number
opposite in cols. of widths, setting prod. in corresponding
col. of prods., and multiplying or dividing by 2, merely
transferring or entirely omitting, as the case may be, to
find the prod. for the other col.   Add the first col. of
widths and the first col. of prods., as before.   For the
cor., multiply the dist. of each interm. from next full sta.
before by the dif. of prods. in second col., one on each
side of interm., added to 9 times [$=Sb$] the dif. of
widths, one on each side of interm. in second col. of
widths, always subtracting the lower from the upper,
using the consequent plus or minus sign, and setting
result in col. of corrections. Since this col. must be
divided by 108, it is unnecessary to consider decimals :

for, assuming the decimals to average half a unit each, 216 interms. must be present in a cut to make the sum of these fractions amount to 108 units, and this is merely equivalent to 1 cu. yd. Moreover, as these corrections have opposite signs, a far greater no. of interms. would be required to make a dif. of a cu. yd. by their decimals. For first interm., add $(101.68-149.04)$, or -47.36, to $(24.8-27.6)\times9$, or -2.8×9, or -25.2. The sum is -72.56: this multiplied by 40 is -2902. For interm. at sta. $4+88$, we have $228.96-221.9+9\times(31.8-31.7)$ $=7.96$. $7.96\times88=700$. Instead of dividing total cor., 1637, by 108, transfer half to results under col. of prods., which have for a divisor 54; but, since these results must be multiplied by 100, move the decimal point of correcting quantity two places to the left.

Concerning the brevity of this method, we have already demonstrated its advantages when applied to single vols. with interm. cross-sections; and of course all the advantages of the parts are collectively the advantage of the whole system, when used to•calculate in one series the entire cut or fill, over the old method of dividing into several series and many single vols. But there are, besides those already discussed, certain merits peculiar to the system as a whole that were not noticeable while considering the single vols. alone. These merits consist rather in avoiding several disadvantages, which in the old method are unavoidable.

Thus, by the old method, the work being computed in separate series and single vols., the end sections of these series and vols., except the first and last cross-sections of cut, are common each to two series or vols., and must each accordingly be included in the calculations of both; so that to estimate a cut or bank containing numerous interms. necessarily the prods. and widths of many cross-sections must be treated twice.

Again, it is evident by the mere mention that a system of numerous separate series and vols., the quantities of each by itself to be added and modified by the common factors of the series or vol., the final results of all again to be combined, necessitates the employment of more time and greater space than does

2

this method of compacting all the quantities in one set of cols., once to be added, and once only to be modified by the factors common to all.

It is likewise as evident that this method of calculating an entire cut or bank in one concise table is more neat, scientific and convenient than the ordinary manner of separate calculation; and that it is therefore also more fit in shape to preserve these calculations.

For these reasons, illustrating the advantages of brevity and convenience, we advocate the practice of this system of calculation, wherein all the full vols., whether pure or containing intermediate cross-sections, are computed together by one rule, and all the intermediates by one correction.

The value of these points of brevity, stated here in the abstract, will be fully appreciated if the above example be examined with this regard. Without using the plan of including interms. hardly any of the advantage of combining vols. can be gained, because scarcely two together of the example have equal length. The first vol., 40 feet long, must be considered alone, its widths and prods. separately added, the former sum mutiplied by Sb, then added to the latter, a subtraction made, and finally the result must be multiplied by 40. In fact, nearly all the work under the cols. of the example as solved by corrections must be done for that single vol. The same is true of the next vol, 60 ft. long. The next two vols. may be calculated together, the following four singly, the next two together, and all the rest singly, making fifteen separate operations to be performed, each containing all the elements of the whole example. Again, the cross-section at sta. $1 + 40$ must be used to find the content of first vol.; it must also be used in another calculation for the second vol. The section of sta. 2 must be used with second vol., and also with the following series. • Sta. 3 need be used but once. With the exception of the first and last stas. always, and in this case of stas. 3 and 6, every section must be used twice. This alone requires 14 extra lines. And, since the calculation thus con-

sists of 15 distinct operations, each requiring for the
sum of its prods. one line, a second for sum of widths,
a third for the subtraction, a fourth for the sum of these
three, and a fifth and sixth to find the prod. of this sum
by length of vol., there must be 15×6, or 90, lines to
accommodate the operations under the cols., instead of
the 7 of our illustration. This is an addition of 83
lines, which with the other 14 make, besides the proba-
ble intervals between the different operations, the last
a consumption of space not time, 97 extra lines for the
calculation. After all this the results of the single vols.
must be added and the sum divided by 108 to find the
cu. yds., and the sum of the results of the different series
must be divided by 54, the last two sums finally combined to make the answer, this summing of results being
equal in space to a col. of our illustration. The relative
amounts of actual figuring is thus perceived.

Of the method by correction the extra col. of widths
and that of prods., although increasing the apparent
bulk of the operation, contain no real extra labor except
the mere writing of the numbers; for, whether to find
the full or half widths of sections for first col., the full
widths must of course be first found, and we have only
to set these in the second col. where required. In cols.
of prods. the full prods., half prods. and blank spaces
occupy positions corresponding with the full and half
widths and blank spaces of the cols. of widths, requiring
a mere transfer from one col. of prods. to the other, or
at most simply a doubling or halving of values.

By the old method as many operations must be performed, with a dist. less than 100 ft. as a factor, as there
are minor vols., 12 in this example: by the correcting
method a dist. less than 100 ft. is used as many times
only as there are interm. stas., 8.

Regarding the differences of widths and prods., used
in the new method, opposed to the sums of these, as
used in the old, let us examine the example. Subtracting one width in the second col. of widths from
the second above, in every instance cancels the
figure of the tens place, leaving only a figure in the
units place, with generally the decimal; while some-

times, as for the interm. $4+88$, where 31.7 is taken from 31.8, both the figures of tens and units places are removed. But to add two widths gives a far more considerable number to use as a factor. Differences in the col. of prods. in almost every case lose the figure of hundreds place, and occasionally of the tens place also, as for interm. $4+88$. The sum of two prods. is a much more troublesome number to use.

CORRECTION OF APPROXIMATION.

THE foregoing method of calculating earthwork is approximate. To find the true contents we use a formula of correction, which is obtained in the following manner. It is well known that the exact content of a vol. of earthwork is obtained by use of the prismoidal formula, whether the ground be a right plane, or the more general hyperbolic paraboloid or warped surface. That is, to find true content add to the sum of end areas four times the area of a cross-section midway bet. ends, and multiply sum by $\frac{1}{6}$ the length of vol. Let

$$\tfrac{1}{2}w(c+Sb)-Sb^2, \ \tfrac{1}{2}w'(c'+Sb)-Sb^2$$

be the end areas of a vol. of earthwork. Then $\frac{1}{2}(w+w')$, $\frac{1}{2}(c+c')$ are evidently the width and centre of mid-section, and its area is

$$\tfrac{1}{4}(w+w')\left(\frac{c+c'}{2}+Sb\right)-Sb^2.$$

Multiply this by 4, add thereto the end areas, multiply all by $\frac{1}{6}D$, and divide by 27, to obtain in cu. yds.

$$True \ vol.=\left(\begin{array}{c}2(wc+w'c')+(wc'+w'c)\\ +3Sb(w+w')-12Sb^2\end{array}\right)\frac{D}{324}.$$

This formula would prove very unwieldy to carry through the calculations for series of vols., especially in the consideration of interm. stas. The same results may be obtained in a simpler manner by using the dif. bet. true and approx. contents as a correction. Subtracting the approx. vol.,

$$(wc+w'c'+Sb(w+w')-4Sb^2)\tfrac{D}{108},$$

from the true, we have for the error or correction

$$\frac{(wc' + w'c - wc - w'c')D}{324} \quad or \quad \frac{(w-w')(c'-c)D}{324}.$$

The advantages of using a separate correcting formula are the following:

If it be desired to make only a hasty approx. estimate, we have a short method by the approx. formula; whereas, were the table of operations constructed upon the prismoidal formula, a great amount of unnecessary work would be unavoidable.

Even if the true contents be required, it is found by trial to be much easier to approximate first and afterward correct than to use the exact formula at once. Thus, to find true content of single vol., we may use the approx. formula and afterward the correction in its simplified form on the right; but to combine these the cor. must be augmented to the difficult shape on the left in order to be taken into the parenthesis. The dif. of labor is still greater in the combinations of vols.

Another very excellent advantage is the facility with which, by means of the correcting formula, it may be at once determined whether a cor. is required at all or not, thus frequently saving much unnecessary labor. Supposing the widths of end-sections to be equal, we see immediately by the correcting formula,

$$(w-w')(c'-c)\tfrac{D}{324},$$

that the cor. is nothing, while the pris. formula requires as much labor for this case as for any other. The cor. would also be nothing if the centres were equal. Again, supposing the centres to vary by one-tenth [0.1], the length of vol. being 100 ft., the widths must vary by 32.4 ft., scarcely a possibility, to make an error of 1 cu. yd.: of a vol. 50 ft. long they must vary 64.8 ft. Hence it is seen that, where the widths or centres of a vol. vary by a few tenths only, the cor. is immaterial. Now, since in any series generally from $\frac{1}{2}$ to $\frac{1}{3}$ of the vols. need no cor., we are able to achieve a correct result within a very small fraction with comparatively little work by the aid of this method of selection, as will be shown in correcting the last example of approximate work.

Another inference to be drawn from the correcting formula is that, where the width and centre of one section are both greater than those of the next, the cor. is a minus quantity; where one measurement is greater and the other less, the cor. is plus : and, since, when one measurement is greater, the other is likely to be so too, the approximation of earthwork by end areas is generally an over-estimate.

We see by the formula that to correct a vol. the dif. of widths, found by a subtraction in one direction, must be multiplied by the dif. of centres, resulting from a subtraction in the opposite direction, this prod. multiplied by the length of vol. and divided by 324. Applying this rule to the second vol. of first example, we have *width* at sta. 2 [27.8] − *width* at sta. 3 [31.5] = −3.7 : *centre* sta. 3 [6.4] − *centre* sta. 2 [5.1] = 1.3. −3.7 × +1.3 × 100 = −481. Set this in an extra col. Treat each vol. in like manner, remembering first and last numbers in col. of widths are half widths. We here represent the col. of corrections of first example. The

PRIS. COR.

```
 —   987
 —   481
 —   208
 —   702
 --  1449
12)3827
  9)319
  3)35.4
 −11.81 Cor.
 3131.37 Approx. con'ts.
 3119.56 True con'ts.
```

sum, −3827, divided by 324 is −11.81 cu. yds. This added to the approx. result yields for the true answer 3119.56 cu. yds. In the second example the dif. of widths of first vol., by subtraction downwards, is −2.7, of centres, by subtraction upwards, is +1.1. −2.7 × +1.1 × 40 = −119. The cor. of second vol. we instantly discover to be inconsiderable, because the dif. of widths is only one tenth. This cor., if calculated, is found to be no more than $\frac{1}{30}$ of a cu. yd. Bet. sta. 4+74 and sta. 5 the corrections are too small to be considered, because its vols. are short and the centres differ by a few tenths only. In this example 8 vols., of the total 17, need no cor., a fact discoverable by the formula without actual labor. The error of the 8 vols. is altogether only ¼ of a cu. yd.

At the expense of $\frac{2}{4}$ of a cu. yd. more, the cor. of 4 other vols. may be dispensed with. So, using the correcting formula, wo may take as little trouble as we please, or, on the other hand, attain as perfect a result as desired.

Instead of dividing the sum of this col., −1504, by 324, move the decimal point two places to the left, and, dividing by 6, place the result with those under col. of prods. as was done with the sum of corrections for interm. stas. The cor. of first example may be likewise treated.

Decimals need not appear in the col. of pris. cor, for the same reason that affects the decimals in the col. of corrections for interms.

CORRECTIONS.

INTERMS	PRISMOIDAL
	− 119
	0
	− 55
	− 52
	− 136
	− 584
	0
	0
	− 230
	− 84
	0
	0
	0
	0
	−184
	0
	− 60
	12)1504
	9)125
	3)14

Correction = − 4.7
Approx. con'ts = 4984.4
True con'ts = 4979.7

LEVEL SECTIONS.

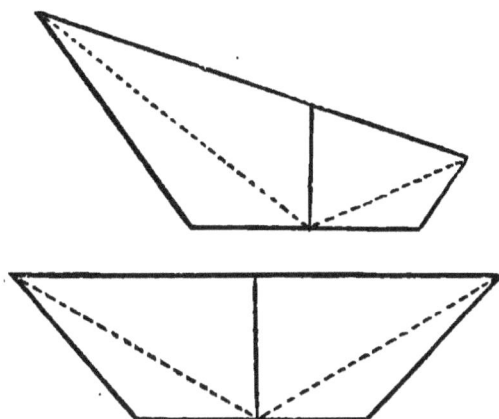

THE surface line of a cross-section may be unbroken at centre stake, as shown in first diagram; and for this case some authors construct especial formulæ, using the symbols of side-heights instead of centre. But such formulæ require greatly more labor than those which use the centre height; and since the last is always known, and the section itself is susceptible of division into the four triangles upon which the formulæ of this paper are based, there is no need to consider this an especial case. But if the surface be level, as frequently occurs in long stretches of river bottom and prairie lands, all the formulæ may be appreciably reduced. Thus for the area, substituting in the formulæ

$$\tfrac{1}{2}w(c+Sb)-Sb^2$$

the value of w in terms of c, $[w=2b+\dfrac{2c}{S},]$ and denoting for convenience the entire road-width by B, we obtain

☞ $$Level\ area = Bc + \frac{c^2}{S},$$

☞ $$Approx.\ vol. = \left(B(c+c') + \frac{1}{S}(c^2+c'^2)\right)\frac{D}{54},$$

$$N \ vols. = \left(B(c + 2c' + \&c.) + \frac{1}{S}(c^2 + 2c'^2 + \&c.) \right)\frac{D}{54},$$

or

$$N \ vols. = \left\{ \begin{array}{l} B(mid\text{-}centres + \frac{1}{2} \ end\text{-}centres) \\ \frac{1}{S}(mid\text{-}centres^2 + \frac{1}{2} \ end\text{-}centres^2) \end{array} \right\} \frac{D}{27},$$

$$Pris. \ cor. = -\tfrac{D}{162 S}(c - c')^2.$$

Let us illustrate by the following example, found on page 99 of Henck's FIELD BOOK, where $B = 28$, $\frac{1}{S} = \frac{3}{2}$, and the centres as recorded in col. headed c, except first and last, which in agreement with the formula are halved:

Sta.	c	c²	(c—c')²
0	1	2	
1	4	16	4
2	7	49	9
3	6	36	1
4	10	100	16
5	7	49	9
6	6	36	1
7	2	8	4
	43	296	44
	28	− 7½	
	1204	288⅔ × ⅔	
		433	
		1204	

9|163700.
3|18188.9
6063. cu. yds.

This table of calculations for level sections is identical with Henck's and others' in cols. headed c and c^2, but differs from all others in last col., which Henck fills with prods. of each pair of consecutive centres. The difference arises from the use here of the correcting formula instead of the exact expression for the whole contents at once : and the advantages attending the use

of the separate cor. are the same for level sections as
for the others. Thus, although the example above is
too simple to show a marked difference, it may readily
be conceived how it is far easier to get the difference
between two consecutive centres, frequently having a
very small remainder and with fewer digits, and square
this, than to find the prod. of the same. In proof of
this we may point out the fact that, while the sum of
the last col. as above is only 44, the sum of the last col.
by Henck's method for the same example is 274, results
indicating directly the comparative bulks of operations
necessary to fill the respective cols.

It is noticed that the cor. embraced by this col. is always a minus quantity for level sections. The divisor
of the sum is 6 times as great as the divisor of the next
col. ; wherefore divide the cor. by 6, setting quotient as
obtained under next col., subtract, divide by S [$=\frac{2}{3}$],
add in the prod. of B [$=28$] by sum of 2d col., and for
cu. yds. divide by 27. When $S=1$, we need simply add
296, $-7\frac{1}{3}$ and 1204. Henck used the value $\frac{2}{3}$ for $\frac{1}{2}$ in
this example because it makes his formula easier.

Intermediate stas. rarely occur on level ground, and
more rarely would they happen to be level also, so the
cor. for interms. need not be reconsidered, though if required it can easily be constructed and applied.

Of course, where but several level-sections occur in a
cut it is not advantageous to alter the mode of calculation or remove them from the general tables, where
they are as correctly treated as the rest; but where
many such sections follow each other consecutively, the
method last discussed is a sensible improvement upon
the former.

IRREGULAR CROSS-SECTIONS.

Over very uneven ground cross-sections of earthwork must necessarily be exceedingly irreg.; and such, it is seen, must interrupt the series of vols. in which they occur, so that by the formulæ already established no entire cut or fill, containing one or more irreg. sections, can be calculated in one operation. As far as we have seen, no formulæ have been constructed for this class of sections, mere hints being given for their convenient computation, as, regarding the diagram, to find areas of trapezoids, $r\ u\ u'\ r'$, $u\ u'\ t'\ t$, &c., and subtract from

their sum the outside triangles, $r\ r'\ r''$, $l\ l'\ l''$, or, dividing into triangles, to solve $r\ r'\ u'$, $l\ l'\ p'$, singly, and in pairs all the other triangles, as $r\ u\ u' + u\ u'\ t = \frac{1}{2} u\ u' \times t'\ r'$, &c., subtracting outside triangles from the sum. We submit the following demonstrations, whose object is to combine vols. bounded by irreg. sections in the same series with the reg.

In the accompanying diagram is represented the right

portion of an irreg. section of one "break," the broken surface line being above the straight line drawn from the surface at centre to top of slope, the latter showing the surface line of an imaginary reg. section of same base, centre and slope.   The area of a whole reg. section is

$$\tfrac{1}{2}w(c+Sb)-Sb^2;$$

of right portion, letting x represent width,

$$\tfrac{1}{2}x(c+Sb)-\tfrac{1}{2}Sb^2.$$

Adding to the latter area the triangle above the reg. portion, we have the area of the irreg. section.   Let m be the height of vertex of break above grade, m' its dist. from centre.   Draw a line from top of slope parallel to base, making a triangle with line to top of centre and with the excess of centre height over slope elev.; also a smaller similar triangle, having for a base part of m denoted by a.   From these triangles we derive

$$x:x-m'::c-r:a\;.\;\;\therefore a=\frac{(x-m')(c-r)}{x}.$$

The part of m below the horizontal line is equal to r; therefore, subtracting $r+a$ from m we have the portion of m included in the upper triangle, which multiplied by $\tfrac{1}{2}x$ is the area of that triangle.

$$Area\;triangle=\tfrac{1}{2}x(m-c)+\tfrac{1}{2}m'(c-r).$$

Adding to the reg. portion of section, we have

$$Area\;right\;section=\tfrac{1}{2}x(m+Sb)+\tfrac{1}{2}m'(c-r)-\tfrac{1}{2}Sb^2.$$

If the break be below instead of above the surface line of reg. section, the

$$Area\;triangle=\tfrac{1}{2}x(c-m)+\tfrac{1}{2}m'(r-c);$$

but in this case its area must be subtracted from the area of reg. section.   Changing the signs, therefore, we have the same formula to add to the formula for reg. section as before.   Hence, in all cases, the formula given above for the area of right-section remains true, whether the break be above or below surface line of reg. section, and, as may also be proven, whether the slope be higher or lower than centre.

Consider the diagram representing a right-section

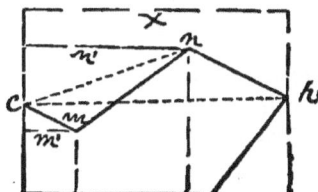

with two breaks. $c\,m\,n\,r$ is the surface line, $c\,r$ the top line of reg. section. Draw $c\,n$. As with last diagram,

$$Area\ triangle\ c\,n\,r = \tfrac{1}{2}x(n-c) + \tfrac{1}{2}n'(c-r),$$

$$Area\ triangle\ c\,m\,n = \tfrac{1}{2}n'(c-m) + \tfrac{1}{2}m'(n-c).$$

If to the formula for reg. part of section the first triangle be added and the second subtracted,

$$Area\ right\ section = \left\{ \begin{array}{c} \tfrac{1}{2}x(n+Sb) + \tfrac{1}{2}m'(c-n) \\ + \tfrac{1}{2}n'(m-r) - \tfrac{1}{2}Sb^2 \end{array} \right\}.$$

This formula represents any form of right-section containing two breaks.

The formula for three breaks, p being the height of third from centre, and p' its dist. therefrom, is similarly found to be

$$\tfrac{1}{2}x(p+Sb) + \tfrac{1}{2}m'(c-n) + \tfrac{1}{2}n'(m-p) + \tfrac{1}{2}p'(n-r) - \tfrac{1}{2}Sb^2\ ;$$

and we see that to find the area of a side-section with any no. of breaks we have only to use the height of last break from centre instead of centre height in the formula for a reg. section, and afterward add one-half the prod. of the dist. of each break from centre by the dif. in height of the breaks next one on each side, the one farther from centre, on surface line, always being subtracted from the one nearer. The same being true of the other side section, we have this

RULE to find the area of an irreg. section :

Multiply one-half each side-width by the height of last break on that side added to Sb. To the sum add one-half the prod. of the dist. of each break from centre by the dif. in elev. of the breaks next one on each side, always subtracting the one farther from centre from the one nearer. Subtract Sb^2.

This method of calculating irreg. sections is similar in principle to the manner of treating interm. stas., before described, the dist. of each break from centre being multiplied by the dif. of heights of breaks next one on each side; and some of the merits of the plan are still perceivable in its latter application. For an irreg. section of five breaks, as illustrated in first diagram, by the rule just given seven prods. must be found, not so difficult as the seven trapezoids of the old method, and instead of finding the areas of the outside triangles and subtracting them, we have only to deduct the constant quantity Sb^2. When combined in series with reg. vols. this subtraction is eliminated, as also the factor Sb, making the process still simpler.

But it is not to the simplification of the treatment of a single section, which must be confessed is comparatively not great in case of the irreg. section, that we now tend, but to the construction of a formula for such sections that may be joined with the formulæ for the reg., in order to avoid the inconvenience of dividing a bank or cut, having one or more irreg. sections, into separated portions, each to be considered alone, and afterward treating singly the vol. on each side of every irreg. section. To obtain this formula, let v represent the left side-width of an irreg. section, m the height of a single break on that side, m' its dist. from centre ; on the right let n and p represent the breaks.

$$\left.\begin{array}{c}\text{Area}\\\text{Irreg.}\\\text{Section}\end{array}\right\} = \left\{\begin{array}{l}Left=\frac{1}{2}v(m+Sb)+\frac{1}{2}m'(c-l)-\frac{1}{2}Sb^2\\ Right=\left\{\begin{array}{l}\frac{1}{2}x(p+Sb)+\frac{1}{2}n'(c-p)\\ +\frac{1}{2}p'((n-r)-\frac{1}{2}Sb^2.\end{array}\right.\end{array}\right\}.$$

Altering the shape of this expression, remembering that $w=v+x$, we have

$$\text{PRODS.}$$
$$Area\ irreg.\ section=\tfrac{1}{2}Sbw+\tfrac{1}{2}\left[\begin{array}{c}vm+xp\\ m'(c-l)\\ n'(c-p)\\ p'(n-r)\end{array}\right]-Sb^2.$$

$$Area\ reg.\ section=\tfrac{1}{2}Sbw+\tfrac{1}{2}\left[\quad wc\quad\right]-Sb^2.$$

It is noticed that the first and last terms of the respective formulæ are identical, and that the second occur in the same col., which may be headed PRODS. Therefore,

$$Area\ any\ section=\tfrac{1}{2}(Sbw+Prods.)-Sb^2,$$

which formula may be carried with perfect facility through all the combinations discussed in former part of this paper, the only distinction bet. reg. and irreg. sections, as represented by their formulæ, being in the definition of the word *Prod.*, which for the former means *the prod. of centre by width,* and for the latter *the sum of prods., each found by multiplying the dist. of a break by dif. of height of two adjacent breaks.* [In case of prods. *vm*, *xp* the same rule applies, since the top of slope may be considered a break distant by *v* or *x*, the next break nearer *m* or *p*, and, no break being beyond, this may be considered zero, whence we have, agreeing with rule, $v(m-0)$, $x(p-0)$.]

Therefore, the rule given for series of vols., as well as the rule for interm. stas., is perfectly applicable to irreg. vols.; and in the table of operations the widths of both fill col. of WIDTHS to be multiplied, together, by common factor *Sb*, the prods. of both fill col. of PRODS. to be added together, and finally the common subtraction, $-Sb^2$, is made once for both, as will shortly be illustrated by an example.

There is another class of cross-sections, which should perhaps be distinguished by the name *defective* or *imperfect*, wherein the entire base of section does not appear, as in the accompanying diagram, where the surface line dips below grade in right-section. Evidently

the portion of section above grade is excavation, and that below, filling; therefore, if the surface line should be considered as it is, we should obtain as a result not the amt. of excavation nor of embankment, but the dif. of their quantities. As this dif. is rarely required alone, excavation and embankment must be considered separately. This is done by regarding the points where surface line crosses grade as breaks, and the surface line to be *l m c n p r*, of which the values of *n* and *p* are

zero. The area of the section is now correctly represented by the formula for an irreg. section of corresponding breaks, viz.,

$$\left\{ \begin{array}{c} \tfrac{1}{2}v(m+Sb)+\tfrac{1}{2}x(p+Sb)+\tfrac{1}{2}m'(c-l) \\ +\tfrac{1}{2}n'(c-p)+\tfrac{1}{2}p'(n-r)-Sb^2 \end{array} \right\}.$$

This formula may be obtained from the section directly,
as well as from all that follow, by means similar to those
used in connection with irreg. sections; and it may be
verified by substituting therein the actual values of the
symbols.

Take another instance where a portion of right-base
does not appear in the section of excavation. Since in

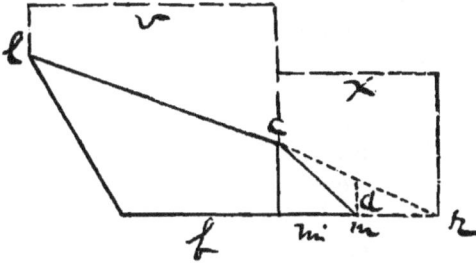

the same series we must always have sections of equal
bases, the base of this section must be prolonged on the
right to the proper dist.; and now the surface line is
considered to be $l\,c\,m\,r$, of which m is the only break,
and m and r zero. Its formula is accordingly

$$\tfrac{1}{2}v(c+Sb)+\tfrac{1}{2}x(m+Sb)+\tfrac{1}{2}m'(c-r)-Sb^2.$$

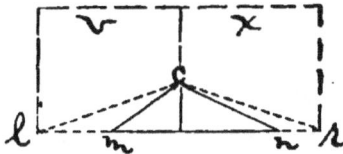

The formula for the next section represented is

$$\tfrac{1}{2}v(m+Sb)+\tfrac{1}{2}x(n+Sb)+\tfrac{1}{2}m'(c-l)+\tfrac{1}{2}n'(c-r)-Sb^2.$$

3

The next section may be regarded as reg. since its

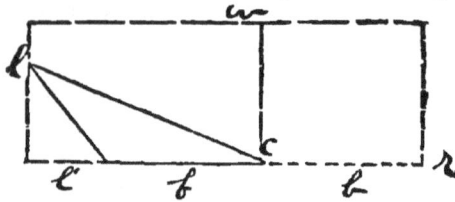

surface line is broken only at centre. Its formula is
$$\tfrac{1}{2}w(c+Sb)-Sb^2.$$

It is certainly not pretended that the formula, as applied to such simple shapes as the last two or the next two following, simplify the same; but it is shown that by means of this formula, as an expression of area, any possible shape of cross-section may be included in the series and thus cause no interruption, which is the great advantage we wish to obtain. Neither does this formula augment to any sensible degree the calculation of simple shapes, it being a mere form serving to direct the values to their proper places in the cols. Thus, for second section above, when in combination, the subtraction, $-Sb^2$, is eliminated, as also Sb; so we have merely to set in col. of widths the entire width, equal to the base of section, and for col. of prods. it is instantly seen that the first two prods. are zero, and the last two, $m'c$, $n'c$. *The width of an imperfect section is the full width of roadway added to the width of whatever sloping the section may have.*

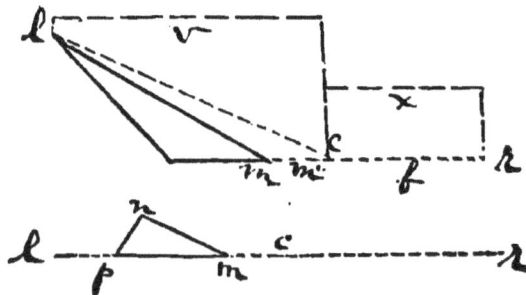

The formulæ for the next two are in order
$$\tfrac{1}{2}v(m+Sb)+\tfrac{1}{2}x(c+Sb)+\tfrac{1}{2}m'(c-l)-Sb^2,$$
$$\left\{ \begin{array}{c} \tfrac{1}{2}v(p+Sb)+\tfrac{1}{2}x(c+Sb)+\tfrac{1}{2}m'(c-n) \\ +\tfrac{1}{2}n'(m-p)+\tfrac{1}{2}p'(n-l)-Sb^2 \end{array} \right\}.$$

The formula for each of the following two sections is

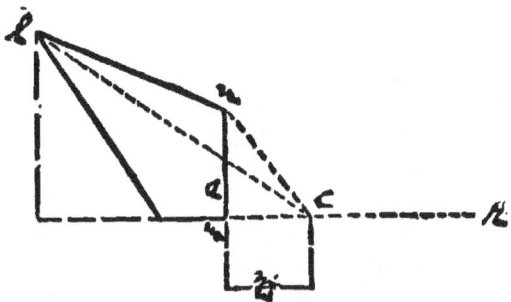

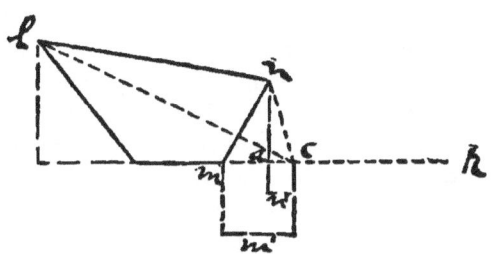

$$\tfrac{1}{2}v(n+Sb)+\tfrac{1}{2}x(c+Sb)+\tfrac{1}{2}m'(c-n)+\tfrac{1}{2}n'(m-l)-Sb^2,$$

which is the general formula for irreg. sections, m being the first break, n the second, *measured on surface line.*

The formula for the next is rather long, but just as

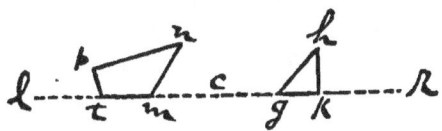

easily constructed by the general law.   Its surface line is $l\,t\,p\,n\,m\,c\,g\,h\,k\,r$.

The last section shown is merely of side trimming,

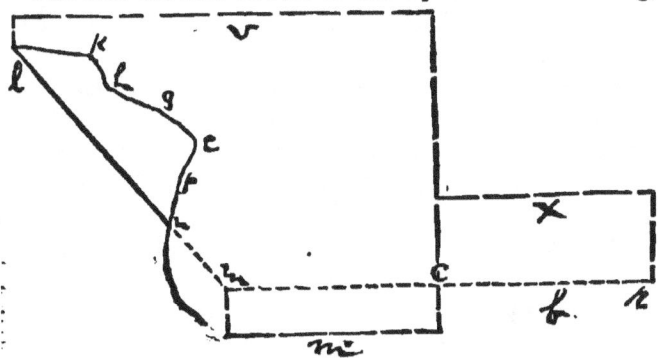

the road-bed not being formed by excavation at all, yet the same rule applies and the same formula represents its area. The surface line is considered to be r c m n p t g h k l, of which the breaks are all in the left section. The first is evidently m, an imaginary pt., whose dist. is known to be the width of half base, and elev. zero; the second, n, the pt. where slope enters material to be removed. The formula need not be re-stated. This case is rare, but is here included to support the assertion that no possible shape of section, within limits of reg. slope and base, lie without the limits of the rule for the calculation of irreg. sections.

We may now calculate approximately in one operation any possible cut or fill. The reg. vols. may be corrected by the formula already given; while the irreg. may also be corrected by a somewhat similar formula, which we shall now proceed to find.

First the general rule may be stated, which is easily susceptible of proof, but for which we shall not yield space here, that *the prismoidal formula may be correctly applied to every vol. bounded by parallel bases, however dissimilar in shape and area, and laterally contained by a surface generated by a straight line moving along the perimeters of the bases as directrices.* This includes all vols. bounded by warped surfaces and surfaces of revolution, which are generated by straight lines, and applies to the majority of shapes occurring in earthwork and masonry.

Irreg. vols. are generally very carelessly treated, on account of the great labor it would incur to calculate the data of mid-sections, then their areas, and afterward apply the primoidal formula. Much of this labor may be removed by using a general formula for such sections, although this plan does not seem to have been hitherto employed. The method, often used, of converting end areas into equal level-section areas, and applying to the resulting vol. the pris. formula, is extremely faulty. In fact, irreg. vols. should be very cautiously calculated, as the ratio of error incurred by using approx. methods is immensely greater than when the same are applied to reg. vols.

In illustration of this, consider the right side of an irreg. vol., shown in diagram with numerical values, m

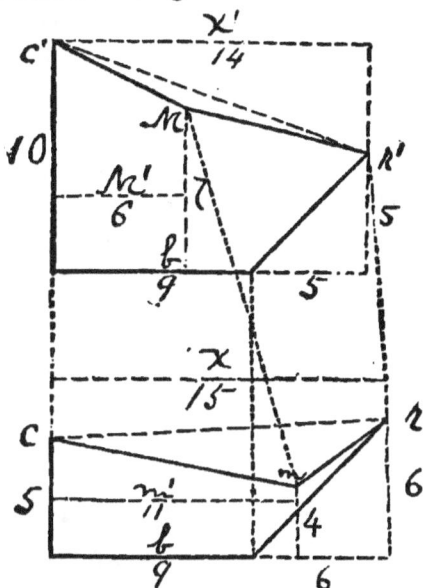

and M being the breaks of the two sections, supposed to represent the features of the same hollow, extending from one section to the other. m and M are therefore considered to be connected by a straight line dividing the two surfaces $m\ c\ c'\ M$, $m\ r\ r'\ M$, one warping from $m\ c$ to $M\ c'$, the other from $m\ r$ to $M\ r'$. The half-base is 9 ft., slope 1 to 1.

$$\text{Area triangle } c\ m\ r = \tfrac{1}{2}x(c-m) + \tfrac{1}{2}m'(r-c),$$

or $$\tfrac{1}{2}.15(5-4) + \tfrac{1}{2}.11(6-5) = 13.$$

$$c'r'M = \tfrac{1}{2}.14(10-7) + \tfrac{1}{2}.6(5-10) = 6.$$

The vol. having these triangles for bases is bounded laterally by a surface generated by a straight line moving along perimeters of bases; hence its content should be calculated by pris. formula. Of mid-section the centre-height is $\frac{c+r'}{2}$, width $\frac{x+x'}{2}$, elev. of break $\frac{M+m}{2}$, dist. $\frac{M'+m'}{2}$, and slope height $\frac{r+r'}{2}$. Substituting these in the formula for area of middle triangle, we have

$$\tfrac{1}{2}\left(\tfrac{x+x'}{2}\right)\left(\tfrac{c+c'}{2} - \tfrac{M+m}{2}\right) + \tfrac{1}{2}\left(\tfrac{M'+m'}{2}\right)\left(\tfrac{r+r'}{2} - \tfrac{c+c'}{2}\right),$$

or numerically, $\quad$ A. mid. tri. $= 6,$

no greater than the area of small-end triangle. This serves as an illustration of a twisted vol., wherein the mid-section and consequently the content is much diminished by the twist. Here we have

$$True\ vol.=(13+6+4\times6)\tfrac{100}{6}=716\tfrac{2}{3}\ cu.\,ft.,$$

$$Approx.\ vol.=(13+6)\tfrac{100}{2}=950\ cu.\,ft.,$$

where it is seen the approx. is a vast excess over the true content: and, since this twisted prism must be subtracted from the vol. bet. reg. sections, the approximation in this instance for the whole irreg. vol. falls short of the true content. Let us see.

$A.\ near\ irreg.\ sect.=\tfrac{1}{2}.15(4+9)+\tfrac{1}{2}.11(5-6)-40.5=51.5.$

$A.\ far\ irreg.\ sect.=\tfrac{1}{2}.14(7+9)+\tfrac{1}{2}.6(10-5)-40.5=86.5.$

$Approx.\ irreg.\ vol.=(51.5+86.5)\tfrac{100}{2}=6900\ cu.\,ft.$

$A.\ near\ reg.\ section=\tfrac{1}{2}.15(5+9)-40.5=64.5.$

$A.\ far\ reg.\ section=\tfrac{1}{2}.14(10+9)-40.5=92.5.$

$Approx.\ reg.\ vol.=(64.5+92.5)\tfrac{100}{2}=7850\ cu.\,ft.$

$Pris.\ cor.\ for\ reg.\ vol.=(10-5)(15-14)\tfrac{100}{12}=+41\tfrac{2}{3}\ cu.\,ft.$

$$True\ reg.\ vol.=\qquad\qquad 7891\tfrac{2}{3}\ \text{"\ "}$$

$$True\ content\ twisted\ prism=\qquad -716\tfrac{2}{3}\ \text{"\ "}$$

$$True\ irreg.\ vol.=265.74\ cu.\,yds.=7175\ \text{"\ "}$$

$$Approx.\ irreg.\ vol.=255.55\ \text{"\ "}\ =6900\ \text{"\ "}$$

We see that the approx. result for the irreg. vol. falls short 10 yds. in a true vol. of 266 cu. yds., only 1½ of which is recovered by application of pris. cor. to the whole reg. vol., while the remaining 8½ yds. is due to the cor. of a mere strip of vol., whose true content is no more than 26½ cu. yds. By this it becomes evident that it is vastly more important to apply the pris. cor. to vols. bet. irreg. sections than bet. reg., the cor. in this case for reg. vol. being but 1½ yds. in 292, whereas the cor. for the irreg. vol. is 10 cu. yds. in a true vol. of 266.

To find the shape of pris. formula as applied to irreg. vols., consider a vol. bounded by irreg. sections, of which c, r, l, v, x, are the centre, right and left slope elevs., left and right side-widths of first, the same

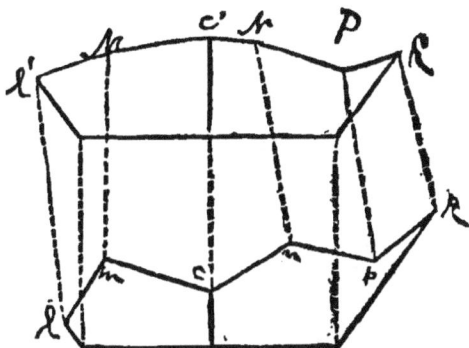

measurements of second correspondingly c', r', l', v', x'. In the first are one break on left, elev. m, dist. m', two breaks on right, elevs. n, p, dists. n', p'; in the other corresponding breaks of alt. M, N, P, dists. M', N', P'. The areas of these by rule are respectively

$$\tfrac{1}{2}v(m+Sb)+\tfrac{1}{2}x(p+Sb)+\tfrac{1}{2}m'(c-l)$$
$$+\tfrac{1}{2}n'(c-p)+\tfrac{1}{2}p'(n-r)-Sb^2,$$

$$\tfrac{1}{2}v'(M+Sb)+\tfrac{1}{2}x'(P+Sb)+\tfrac{1}{2}M'(c'-l')$$
$$+\tfrac{1}{2}N'(c'-P)+\tfrac{1}{2}P'(N-r')-Sb^2.$$

The corresponding values of mid-section are found by dividing the sum of similar measurements of end-sections by 2. Its area is, therefore,

$$\tfrac{1}{4}(v+v')(\tfrac{m+M}{2}+Sb)+\tfrac{1}{4}(x+x')(\tfrac{p+P}{2}+Sb)$$
$$+\tfrac{1}{4}(m'+M')(\tfrac{c+c'}{2}-\tfrac{l+l'}{2})+\tfrac{1}{4}(n'+N')(\tfrac{c+c'}{2}-\tfrac{p+P}{2})$$
$$+\tfrac{1}{4}(p'+P')(\tfrac{n+N}{2}-\tfrac{r+r'}{2})-Sb^2.$$

Multiply mid-section by 4, add thereto the sum of end areas, and multiply by $\tfrac{1}{6}$ length, to find true vol. From this subtract the approx. amt., found by multiplying sum of end areas by one half length. The remainder is the pris. cor., being in cu. yds.

$$\left\{ \begin{array}{c} (v-v')(M-m)+(x-x')(P-p) \\ +(m'-M')((c'-l')-(c-l))+(n'-N')((c'-P)-(c-p)) \\ +(p'-P')((N-r')-(n-r)) \end{array} \right\} \frac{D}{324}.$$

This is more conveniently expressed in words. The first term is the dif. of corresponding side-widths multiplied by the dif., taken inversely, of corresponding last breaks on that side. The second term is similar. In the third term one factor is the dif. of dists. of corresponding breaks from respective centres, and the other the dif., taken inversely, of the quantities, which are multiplied by those dists. to find areas of respective sections. In the formulæ for the areas of first and second sections we find the terms, $\frac{1}{2} m' (c-l)$, $\frac{1}{2} M' (c'-l')$, which are used in the approx. calculation, m' and M' being placed in their proper col., and $c-l$, $c'-l'$, in another; so that to apply pris. cor. we have only to find a dif. by subtraction downwards in one col. and multiply it by a dif. found by subtraction upward in the other col., a method precisely similar to that employed for reg. sections. The third term is a type of all that follow. Therefore we may construct this

RULE OF PRIS. COR. for vols. bet. irreg sections.

Multiply the dif. bet. each two corresponding side-widths by the dif., taken inversely, of the elevs. of last breaks on that side. To these prods. add the prod. of the dif. of dists. from respective centres of each two corresponding breaks by the dif., taken inversely, of the respective factors used with these dists. to find the areas of respective sections. Multiply the sum by $\frac{1}{3}$ dist. bet. sections and divide by 27.

This rule requires that the same no. of breaks be present in each end section of a vol., and that the breaks of one end be connected with the corresponding breaks of the other by straight lines. This arrangement is true of every vol., although the breaks may not appear in one section. In the example just considered it is perfectly obvious, where the three breaks of one connect with the three of the other; but sometimes one end section has more apparent breaks than the other. Thus, in proceeding from reg. to irreg. ground the last reg. section is succeeded by an irreg. section of one or more breaks, which do not, however, originate in that section itself, but gradually develop themselves from certain points on the profile of reg. section. These points must be ascertained in order to make a correct computation, and they represent the breaks of the reg. section corresponding to the actual breaks of the irreg. That these ridges and hollows do extend from one section to the other is self-evident. The irreg. section should not represent mere local features, since its area and shape affect the vol. all the way to next section on each side. The ridges and hollows may be faint, but, if visible at all in one cross-section, their general course may be traced, and the pt. of intersection with reg. section nearly determined.

The neglect of these fading ends is the source of great error, although at first thought it would be considered perfectly immaterial to the result where the ridges and hollows might happen to fade on the surface line of reg. section. We shall give space to one illustration.

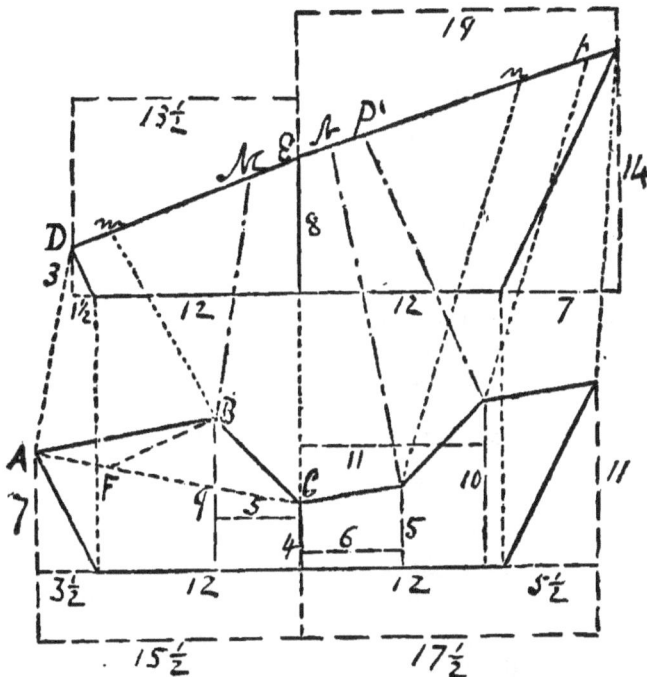

The vol. is bounded by a reg. and an irreg. section, whose measurements are marked in the diagram. M, N, P, are the pts. on reg. section, to which the breaks of irreg. section are supposed to extend in one instance; m, n, p, the pts., to which the breaks are supposed in another instance to extend. The approx. formula for content is

$$Approx.\ vol. = (prods. + Sb \times widths - 4Sb^2)\tfrac{D}{108};$$

applying which to the example we have

W		C		PRODS.
$\begin{cases} 32\frac{1}{2} \\ 15\frac{1}{2} \\ 17\frac{1}{2} \end{cases}$	× × ×	8 9 10	= = =	260 139.5 175.
65.5 5	×	(4 − 7)	=	−15.
24 6	×	(4 − 10)	=	−36.
2620 11	×	(5 − 11)	=	−66.
1310				457.5
1572.0				1572.
				−1152.
				9)877.5
				12)9750.

Approx. vol. in cu. yds. = $\overline{812.5}$

Now if the ridges and hollows represented in irreg. section vanish at the pts. *M, N, P*, of the reg., these pts. may be considered corresponding breaks, and all the conditions for the application of pris. cor. are present. The distances of these pts. are noted in field, the heights being readily deduced, as by the formula $h = a + \frac{d(b-a)}{l}$, where h is the height of intermediate pt., a the elev. of first end of line, b the elev. of second end, d the dist of pt. from first end, and l the horizontal length of line. Thus, if *M', N', P'*, are 3, 2, 4, respectively,

$$M = 8 + \tfrac{3(3-8)}{13\frac{1}{2}} = 6.89, \quad N = 8 + \tfrac{2(14-8)}{18} = 8.63,$$

$$P = 8 + \tfrac{4(14-8)}{19} = 9.26.$$

Applying the pris. cor.,

$$
\begin{aligned}
(15\tfrac{1}{2} - 13\tfrac{1}{2})(6.89 - 9) &= &- 4.22 \\
(19 - 17\tfrac{1}{2})(10 - 9.26) &= &+ 1.11 \\
(5 - 3)((8 - 3) - (4 - 7)) &= &+ 16.00 \\
(6 - 2)((8 - 9.26) - (4 - 10)) &= &+ 18.96 \\
(11 - 4)((8.63 - 14) - (5 - 11)) &= &+ 4.41 \\
& & 324)\overline{3626.}
\end{aligned}
$$

pris. cor. = + 11.2 *cu. yds.*

When the breaks represented in irreg. section extend to the pts. *M, N, P*, of the reg., the true content bet. is

$$812.5 + 11.2 = 823.7 \ cu. \ yds.$$

Let it now be supposed that these breaks extend to

the pts. m, n, p, distant 11, 13 and 15, and of elev. 3.9, 12.1, 13.4. Apply cor.

$$
\begin{aligned}
(15\tfrac{1}{2}-13\tfrac{1}{2})(3.93-9)= & \quad -10.14 \\
(19-17\tfrac{1}{2})(10-13.36)= & \quad -15.04 \\
(11-5)((4-7)-(8-3))= & \quad -48.00 \\
(13-6)((4-10)-(8-13.36))= & \quad -\ 4.48 \\
(17-11)((5-11)-(12.1-14))= & \quad -24.6 \\
324) & \overline{-10226.}
\end{aligned}
$$

$$\textit{pris. cor. in cu. yds.} = \quad -\ 31.56$$

When the breaks of the vol. extend to the pts. m, n, p, of the reg. section, the true content is

$$812.5 - 31.56 = 780.94 \ \textit{cu. yds.}$$

Therefore, if M, N, P, be the vanishing extremities of these breaks and remain unnoted, the approx. result falls short of the true content by 11.2 cu. yds. If, otherwise, m, n, p, be the fading ends, the approx. calculation is an over-estimate by 31.56 cu. yds. The total effect on the vol. occasioned by moving the extremities of breaks from the pts. M, N, P, to the pts. m, n, p, is a diminution of 43 cu. yds.; and by other arrangement of the pts. this may be made more than 50 cu. yds.

It may be easily determined at a glance whether the effect of shifting the vanishing extremities be great or little upon the vol. In left irreg. section conceive a line to be drawn from top of slope to top of centre, and a warped surface to extend from this to the left surface line of reg. section. This divides the triangular warped-faced wedge $ABCDE$ from the main vol.; and the latter is evidently not affected by the shifting of M along DE. But the triangular wedge is affected to the following extent. If B be moved along a line BF parallel to DE a certain dist., and the consequent increment or decrement to the wedge denoted by h, M remaining fixed, then the vanishing end M be moved along DE the same dist. in the same direction, B remaining fixed, the second consequent increment or decrement is equal exactly to $\tfrac{1}{2}h$. Therefore, to determine whether the neglect of noting the vanishing end, M, of a ridge or hollow be or be not serious, it is only necessary to notice the inclination of the lines AC, DE. If they are parallel, change of position of M does not alter the content of wedge ; but, if they are much inclined, the position of M is of great importance as already shown. Thus, to apply the principle, it is seen at once that the effect of shifting M in left section is far greater than can be produced by altering the positions of N and P in the right.

It is thus apparent that to obtain all the data for a correct computation of an irreg. vol., both ends of every ridge and hollow must be noted, whether it fade out or not ; that is, each end-section must have the same no. of corresponding pts. connected by straight lines. To take these notes in the field is simple. If we find in taking cross-sections two adjacent sections have three breaks each, corresponding, no remarks may be made ; and in applying pris. cor. it will be taken for granted that these are respectively connected by straight lines, the nearest to the centre of one section with the nearest of the other, the farthest with the farthest, and middle with middle, since they do not intersect bet. cross-sections. If one have four while the other has three, we have only to find the position of fading end of vanishing break in section representing but three : its dist. from centre of this section is placed in col. of remarks opposite this section. When the pris. cor. is applied, the four breaks of one section are connected with the three real breaks and the fading pt. of the other, according to their dists. from respective centres, as before. In all cases, after noticing what breaks of one connect with what breaks of the next, only the vanishing extremities of the other breaks need be noted, obtaining thus always an equal no. of pts. in each, which to connect according to the dists. from their respective centres. This will be fully illustrated in the following example.

The first col. of this extract from field book contains the nos. of stas. and interms, the second the centre cutting or filling, preceded, if the latter, by minus sign. The third col. has the elevs. and dists. of left slopes, with the heights and dists. of interm. breaks, ranged in their natural order. The fourth contains the right dists. and elevs. similarly arranged. The elevs. in all are marked above the dists. of same pts. **The full stas. are** 100 ft. apart, the width of roadway 30 ft., the rate of slope for the cutting is $\frac{1}{2}$ hor. to 1 vert., represented here by $\frac{vert.}{hor.} = 2 = S$. The slope of the filling is $1\frac{1}{2}$ to 1; but we propose to calculate the excavation only, wherefore the left-widths of such sections as dip below grade on that side are considered to be 15 ft. In the col. headed REMARKS are placed the data concerning the ridges and hollows. For instance, sta. 1 being irreg. and sta. 1+50 reg., the fading extremities of the ridges and hollows profiled in the former must be determined in the latter. The left break is found to extend to centre of sta. 1+50, and it is so marked on a little line extending to the left from a vertical line representing centre line. The right break extends to the top of slope, and is therefore recorded as 16.2 on the right of vert. line (these vert. lines appear rather contracted in the print.) Both these records are placed at the top of vert. line, leaving the lower end for possible notes connected with the next following section. But here none is needed as the sections fully correspond as they are. Sta. 3+20 is irreg., therefore the corresponding pts. of sta. 3 must be noted. The left break extends to left slope, being a grade-line bet. the sections; the right break extends to centre, whose dist. is zero. Stas. 3+20 and 4 do not correspond. A pt. on sta. 4 corresponding with left break of sta. 3+20 must be found, and a pt. on right of sta. 3+20 corresponding with second break on right of sta. 4. The two right breaks of sta. 4 are found to merge into the one of sta. 3+20; the dist. 10 is therefore noted in the remarks. One of the two breaks of sta. 4 disappears at centre of sta. 5, and is so noted. The other runs through sta. 5 and disappears at sta. 6 at a dist. of 9 ft. from centre. Bet. stas. 6 and 7 a grade line runs from left slope of former to centre of latter; these pts. are noted. This grade line continues to stas. 8 and 9, appearing in these as breaks; it must be noted in sta. 7 as a break corresponding with that on right of sta. 8. No natural example would be likely to have so many breaks vanishing and originating in so small a dist.; it is only made so here to include all the cases.

Field Book.

Sta.	Centre.	Left.	Right.	Remarks.
1	2			
+50	2			
2	3			
3	5			
+20	5			
4	6			
5	7			
6	6			
7	0			
8	−5			
9	−4			

TABLE OF OPERATIONS.

STA.	WIDTHS.		CEN.	PRODUCTS.		CORRECTIONS. INTERMS.	PRISMOIDAL.	BREAKS.	
1		16.	1		16.			15.8 ×	2.0
"		16.5	4		66.			16.2 ×	2.4
"	L	11.	0		0.	+4.4		0 ×	0.4
"	R	9.	÷ 1		— 9.	—4.32	0	16.2 × —	0.4
+ 50	16.	32.	2	32.	64.	—1060		15.0 ×	0.0
2	32.4	32.4	3	97.2	97.2		0	19.0 ×	5.0
3	17.	34.	5	85.	170.		— 320	15.0 ×	5.0
+ 20	7.5	15.	0	0.	0.	—1580	1140	0.0 × —	3.0
"	11.	22.	6	66.	132.			22.0 ×	6.0
"	L	8.	5	20.	40.			10.0 × —	1.0
"	R	10.	— 9	— 45.	— 90.			10.0 × —	8.0
4	15.	15.	6	90.	90.		— 240	15.0 ×	0.0
"	23.	23.	9	207.	207.			15.0 ×	6.0
"	R	11.	— 3	— 33.	— 33.			24.0 ×	8.0
"	R	15.	— 9	—135.	—135.			0.0 × —	1.0
5	17.	17.	7	119.	119.		1500	12.0 × —	11.0
"	24.	24.	8	192.	192.			15.0 ×	6.0
"	R	12.	—11	—132.	—132.			27.0 ×	12.0
6	42.	42.	6	252.	252.		—3500	9.0 × —	18.0
7	45.	45.	0	0.	0.		—7200	15.0 ×	0.0
8	15.	15.	0	0.	0.			27.0 ×	6.0
"	29.	29.	0	0.	0.			15.0 ×	6.0
"	R	5.	—28	—140.	—140.		—1000	15.0 ×	0.0
9	7.5	15.	0	0.	0.			30.0 ×	0.0
"	12.5	25.	0	0.	0.			0.0 ×	0.0
"	R	11.	—20	—110.	—220.		—4800	15.0 ×	0.0
	313.9			565.2		—2640	—14420	30.0 ×	0.0
								0.0 × —	30.0

Prismoidal working figures (bracketed): 0, —5, 60, 0, —3, —7, +2, +5 { 57×20, —3×80 }; —2, +1, +22, —6 { +15×100 }; —2, —12, —21, 0, —90, +18, 0, 0, —48 { —35, —72 }.

Breaks side-notes: Sta. 1+50 for Sta. 1. — Sta. 3 for Sta. 3+20. — Sta. 3+20 for Sta. 4. 3+20. — Sta. 5 for Sta. 4. — Sta. 6 for Sta. 5. — Sta. 6 for Sta. 7. — Sta. 7 for Sta. 8. — Sta. 8.

$Sb \times widths$ .... 9417.

$-2Sb^2n$ .... —7200.

Cor. for Int's ÷ 200. — 13.2

Pris. Cor. ÷ 600 ... — 24.033

9)274496.67

6)30499.63

5083.27 cu. yds.

For the approx. calculation of this piece of excavation the notes in col. of remarks are of course not needed, since they do not affect the areas of the sections. According to the approx. rule on page 13, construct a table of cols., the 1st for stas., the 2d and 3d for widths, the next for centres, two more for prods., and another for the cor. for interms. Finally add one col. for pris. cor. and one for extra notes. In 2d col. of widths set the full widths of reg. sections; and place the centres of these opposite in col. of centres. For irreg. sections place in col. of widths the side-widths separately, and in col. of centres the factors used with these side-widths to obtain prods., namely, the heights of last breaks ; also in 2d col. of widths place the dist. of each break, and in col. of centres the factor used with this, namely, the height of break next nearer centre minus the height of break next farther. Thus for first section we have directly from field-book, 16×1, 16.5×4, $11 \times (2-2)$, $9 \times (2-3)$; for sta. $3+20$, 15×0, 22×6, $8 \times (5-0)$, $10 \times (5-14)$; for the next, 15×6, 23×9, $11 \times (6-9)$, $15 \times (7-16)$; for the last, 15×0, 25×0, $11 \times (0-20)$. These prods. form the 2d col. of prods. For the first cols. of widths and prods. merely transfer values in 2d cols. according to rule, that is, full widths and prods. for mid-stas., half-widths and prods. for two end stas. of work, with this exception, that half the widths and prods. of last interm. of a vol. must be used in place of half the widths and prods. of the full sta. next before. .But this principle has been already discussed. Since the first sta. is one of the two end stas. and, therefore, only half its value could be used, and since this half must be omitted in favor of the half prods. and widths of the interm. following, no transfer is made for sta. 1. For sta. $1+50$ transfer half the width and half the prod. For sta. 2 transfer entire values. Transfer the half values of sta. 3, since it is followed by an interm. For the interm. transfer half the width and half the prods. For all the rest except the last transfer entire values, for the last half values. In first col. of widths are placed the letters L and R opposite break dists. of sections. These letters serve two purposes.

First, they occupy all space of first col. except what should be filled with transferred quantities, thus preventing the possible mistake of transferring the break dists. also to first col. of widths. Their special use is to mark corresponding breaks while applying pris. cor.

To apply cor. for interms. to first interm. we use the formula

$$(Sb(w-w') + P - P')\tfrac{D}{10b},$$

where w and P are the width and prods. of next section before, or sta. 1, w', P', the same of next section following, or sta. 2, $Sb = 2 \times 15 = 30$, $D = 50$; the factor $\tfrac{1}{10b}$ will be used at foot of col. So we have

$$0.1 \times 30 + (-24.2) = -21.2 \ ; \ -21.2 \times 50 = -1060 :$$

for Sta. $3+20$, $-4 \times 30 + 41 = -79$; $-79 \times 20 = -1580$.

To apply pris. cor. to vol. bet. stas. 2 and 3, which is reg., multiply dif. of widths of end-sections by inverse dif. of centres, and multiply by length of vol. $-1.6 \times 2 \times 100 = -320$. The widths of vol. above differ by a few tenths only, wherefore its cor. may be safely neglected. For the cor. of irreg. vol. bet. stas. 8 and 9, we have arranged before us in perfect order the values considered in formula of pris. cor. The rule requires that the dif. of corresponding widths be multiplied by inverse dif. of last breaks, or of centres, if no break be on that side; also that the dif. of corresponding break dists. be multiplied by inverse dif. of respective factors, viz , those in col. headed ᴄᴇɴᴛʀᴇ, used with these dists. to find areas of respective sections. In applying this rule advantage may be taken of the fact, already mentioned in connection with reg. vols., that, if corresponding values of two sections differ by a few tenths only, the cor. for the vol. bet., so far as affected by those values, is immaterial. Thus, in the last vol. the left side-widths are equal, whence their dif. is zero, and the cor., as affected by them, is zero. Again, although the right-widths differ, their respective factors are equal, and the cor., as affected by them, nothing. But for the third corresponding values the cor. is considerable, being $(11-5)(-28-(-20)) = -48$, giving for the vol.

4

−4800 to set in last col. Now, in a natural example of irreg. earthwork, the majority of vols, would have their corresponding values thus conveniently arranged in the table, making the cor. a simple matter. But here we have purposely considered numerous short breaks, making vanishing extremities at nearly every sta.; so the last is the only irreg. vol., to which the cor. can be directly applied. For the others the notes taken in col. of remarks must be consulted. For instance, the reg. sta. 1+50 must be treated as if irreg. to correspond with sta. 1, the dists. of its breaks being given and their heights found. So its factors would be, to correspond with those in cols. of widths and centres, belonging to sta. 1, the following; 15.8×2, 16.2×2.4, 0×0.4, 16.2×-0.4; that is, the sectional notes must be considered as,

Left   Centre   Right

$\frac{15.8}{2}, \frac{0}{0}, \qquad 2, \qquad \frac{16.2}{2.4}, \frac{16.2}{0.4}.$

But it is unnecessary to place the sectional notes in this form, as the proper corresponding factors may be taken directly from field-book. These factors being obtained, we proceed as before, using them in connection with those belonging to sta. 1 and presented in the table. Thus, since 16 and 15.8 vary by so little, zero may be accepted as the cor., so far as these values affect it. The cor. for next values is also zero. Then we have $(11-0)(0.4-0) = +4.4$, $(16.2-9)(-1-(-0.4))$ $= -4.32$. Sta. 3 must be arranged for sta. 3+20, as shown in supplementary col. headed Breaks. Although stas. 3+20 and 4 are both irreg., their breaks do not correspond, wherefore the left of sta. 4 must be arranged as if containing a break, and right of sta. 3+20 as if containing two. The left of sta. 4 then becomes as shown in col. of breaks, where the right of sta. 3+20 is also seen. For the differences, from the left-width of sta. 3+20 as in table subtract the left width of sta. 4 as in col. of breaks; the dif., and consequently the cor., is zero; then from left break dist., 8, of sta. 3+20 subtract left break dist., 15, of sta. 4, and multiply dif. by inverse dif. of their factors. For

the right-side measurements, use those in the table for
sta. 4, and those in col. of breaks for sta. 3 + 20. The
right of sta. 5 must be re-arranged for sta. 4, and
the right of sta. 6 for sta. 5. The fading end at
sta. 6 is 9 ft. from centre; therefore $h = 6 + \frac{\circ12\frac{1}{2} - 6}{7}$
= 12, and the notes of the right section are, 6, $1\frac{1}{3}\frac{3}{3}$,
$\frac{3}{3}\frac{3}{3}$. The sections at stas. 6 and 7, although reg., include
an irreg. vol., containing a hollow bet. its centre
and left slope lines. Both sections must be arranged as
if irreg. that this hollow may not be neglected. These
arrangements are both shown in col. of breaks, and must
be considered together while applying the cor. Sta. 7
must also be arranged to correspond with sta. 8. Stas.
8 and 9 correspond.

By the approx. rule [see pages 32 and 13] multiply
sum of widths by $Sb[=30]$, add prod. to sum of prods.,
and subtract $2S'^2n$, n being no. of full vols., found always
by subtracting the no. of first full sta. from the no. of
last full sta. Next the rule directs to multiply by 100
and divide by 54; but, since the cor. for interms. and
the pris. cor. must be respectively divided by 2×54 and
6×54, place $\frac{1}{2}$ the first and $\frac{1}{6}$ the second under col. of
prods., moving the decimal pt. of each two places to the
left, because the quantities with which they combine
will be multiplied by 100 and thus carry the decimal pt.
of these corrections back to the proper place.

It has been stated that a natural example would not
be likely to have so many vanishing ends of breaks, and
would, therefore, have its values arranged at once in
convenient form for the application of pris. cor. When
these vanishing ends do occur, the work of rearranging
the sections containing fading ends is necessary to the
true calculation by any method, as the demonstrations
on this point were independent of method. But, having
found them, there can be no more convenient plan of
applying the pris. formula than this method of cor. by
using differences of corresponding measurements, these
being ranged before us in proper order. By this general
method of calculating irreg. sections and including
irreg. vols. in the series we have succeeded in calculating in one operation ten consecutive vols., of which four

are of minor length and eight are irreg. Without the method of including irreg. vols., every vol: of the example above must be calculated singly, every cross-section but first and last be used twice, and the mid-section of every irreg. vol. measured and calculated. In fact, the inclusion of irreg. vols. in series serves, in the same manner as the inclusion of minor vols., to abbreviate the computation of earthwork, and chiefly to remove interruption to the continuous calculation of a long series of vols. in one operation. By including both minor and irreg. vols. all interruption is removed from the thorough and exact calculation in one operation of any possible cut or fill; and the entire process is governed by but four rules. These are, the rule for the approx. amount with its auxiliary rule for intermediates, here remembering the distinction between *Prod.* as referred to reg. and to irreg. sections, the formula of pris. cor. for reg. vols., and the rule of pris. cor. for irreg. vols.

END VOLUMES.

THE method preceding embraces all of a cutting or bank bet. end full stas., leaving to be calculated the vols. tapering from last full stas. to grade lines.

If there be a cross-section bet. last full sta. and grade line, the approx. content bet. this section and last full sta. may be found by the formula

$$(wc + w'c' + Sb(w+w') - 4Sb^2)_{\frac{D}{108}},$$

if reg., or in any case by the general formula

$$(prods. + Sb \times widths - 4Sb^2)_{\frac{D}{108}},$$

and the pris. cor. by the rule.

If the grade line happen to be unbroken and perpendicular to centre line, it may be regarded as a reg. section, whose centre and slope elevs. are all zero, and width the width of road-way, and the rules of approximation and correction accordingly applied.

But, if the grade line, represented in the diagram bet. pts. E, F, G, be broken or be not perpendicular to cen-

tre line, for an approx. method suppose a plane passed through each side slope line of last section and centre grade pt., as rF, lF in diagram. These divide end vol.

into three approx. pyramids, from which is obtained this

Rule for approx. calculation of vol. bet. end-section, this being reg., and grade line.

Multiply area last section by dist. to centre grade pt.: add the prod. of $\frac{1}{2}$ b by sum of prods. of each slope height by dist. on that side to grade. Divide by 81.

Applying this rule to the example, we gain as a result 62.59 cu. yds. This is probably the simplest possible rule.

Otherwise, by passing a vertical plane through each side grade pt. and centre of last section, the end vol. is again divided into three approx. pyramids, the side sections being bases of outer ones and side dists. to grade their heights, and of middle pyramid the centre elev. of last section being alt. and all portion of road-bed, unoccupied by side vols., the base. Then

$$Content\ left\ pyramid = \tfrac{1}{6}Ev(c+S\hbar) - \tfrac{1}{6}S\hbar^2 E.$$
$$Content\ right\ pyramid = \tfrac{1}{6}Gx(c+S\hbar) - \tfrac{1}{6}S\hbar^2 G.$$
$$Content\ middle\ pyramid = \tfrac{1}{3}Fbc.$$

$$\therefore\ End\ vol = ((c+S\hbar)(Ev+Gx) - S\hbar^2(E+G) + FBc)_{\tfrac{1}{162}},$$

B representing full width of roadway. By this formula we obtain for content of example

$$\left\{ \begin{array}{l} (22 \times 18 + 17 \times 10)(8+12) \\ -144(18+10) + 8 \times 15 \times 24 \end{array} \right\}_{v \times \frac{1}{6} \times 3} = 62.81.$$

The average of results of these two methods, 62.7, is always the exact content; but, as in this case, where the surface is not much warped, either result is sufficiently near the truth.

Another method in use is to divide end vol. into four parts by three vertical planes, one through centre line and one through each edge of base, making two outside pyramidal vols., and two inner vols. having quadrilateral bases and grade lines for edges. This requires much more work than either above approx. method. The result of this method for the present example is 62.53 cu. yds.

If the surface line of last section be slightly irreg.,

any of the foregoing rules of approximation may be applied, the latter preferred since longitudinal planes are apt to cross fewer breaks than the diagonals. But, if the breaks be important, making the grade line irreg. as well as the surface line of section, the vol. should be divided by a diagonal plane through each warped surface.

ESTIMATION OF FINISHED WORK.

THE completed work seldom coincides with the prescribed lines, except in ordinary homogeneous earth-
cuts, and in embankments generally. When this is so,
of course, the new cross-sections of actual excavation
must be calculated; and, since these have no measurements in common but are irregular all around their perimeters, they cannot advantageously be computed in
series, so that not only all the methods of brevity of the
preliminary calculation are lost to the more important
estimation of the final quantities but also the prismoidal
correction. For instance, in the accompanying diagram

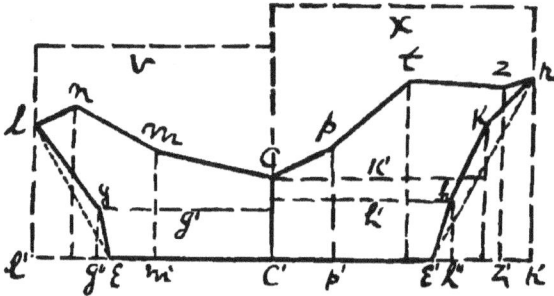

the section of excavation, limited by the full lines, must
be estimated in the ordinary rude way by computing
the trapezoids $n\,l'$, $n\,m'$, $c\,m'$, $c\,p'$, $t\,p'$, $t\,z'$, $r\,z'$, and subtracting from their sum the trapezoids $k\,r'$, $k\,h''$, $g\,l'$, and
the triangles $g\,g''\,e$, $h\,h''\,e'$.

We propose to avoid this labor and inaccuracy by
considering the lower outline, $l\,g\,e\,e'\,h\,k\,r$, of the section of excavation as the surface line of a section
having a regular base, $e\,e'$, and slopes, $e\,l$, $e'\,r$; then
calculating in one series by the method already explained, using the cor. for interms. and the pris. cor.,
the amount of material through the whole cut remaining to be removed in order to expose the true slope
and base lines; and finally subtracting this from the
preliminary result to find the true content of the mass
removed. The advantages of this plan are, briefly, that

the necessity of noticing the measurements of the upper
part of perimeter is escaped, these being considered in
the preliminary calculation, and also that the remaining
measurements may be arranged in a series similar to the
original, yielding a result with all the correctness of the
prismoidal formula.

The surface line of the section of remaining excavation, represented in diagram, is $l\ g\ e\ c'\ e\ h\ k\ r$, of which
the breaks on left are in order e and g, on right e', h and
k. By the rule its area is

$$\tfrac{1}{2}v(g+Sb)+\tfrac{1}{2}x(k+Sb)+\tfrac{1}{2}b(c'-g)+\tfrac{1}{2}g'(e-l)$$
$$+\tfrac{1}{2}b(c'-h)+\tfrac{1}{2}h'(e'-k)+\tfrac{1}{2}k'(h-r).$$

This formula is very simple, especially in combination,
e, c' and e' being each zero.

Sometimes the lower part of perimeter transgresses
the proposed lines of slope, as in the diagram. The

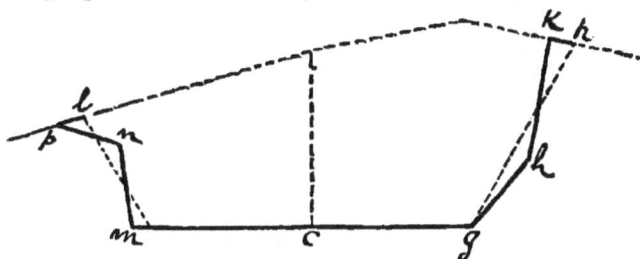

rule still applies to finding the area of such a section,
its value being minus in this instance, as it should be,
because a minus amt. of material must be removed to
bring the work to the proper slope lines; and this
minus amt., representing the work remaining to be
done, when subtracted from the original amt. prescribed, leaves an amt. greater than that originally proposed,
as the case requires. Occasionally the road-bed is
excavated deeper than it should be; then c, m and g
become minus (in case it be determined to consider the
unnecessary work.) But to every possible shape the
same rule applies, yielding positive or negative quantities as it should be : so no allowance need be made for
certain positions of perimeter, since the rule takes care
of itself entirely, without other superintendence than to
follow it exactly, as may be verified by other methods

of calculating the area. Of the section last sketched,
the first break on right is at g, the second at h, the third
at k, where the perimeter reaches the original surface
nearer centre than is the top of prescribed slope. On
the left the first break is at m, where the road-bed has
been cut too wide, the second at n, the third at p, where
the perimeter reaches the surface beyond the slope
stake at l. Accordingly, the surface line of section to
be calculated is $l\ p\ n\ m\ c\ g\ h\ k\ r$, the side-widths of new
sections being always the same as those of the old, and
its area is

$$\tfrac{1}{2}v(p+Sb)+\tfrac{1}{2}x(k+Sb)+\tfrac{1}{2}m'(c-n)+\tfrac{1}{2}n'(m-p)$$
$$+\tfrac{1}{2}p'(n-l)+\tfrac{1}{2}g'(c-h)+\tfrac{1}{2}h'(g-k)+\tfrac{1}{2}k'(h-r).$$

Of course, these sections can be arranged in a series
like the original. The notes for the pris. cor. are more
easily taken on the finished work than on the original
surface, because the outlines are sharper, and the same
features are likely to extend long distances, being occasioned by the dividing lines between strata of various
hardness.

The finished work may be divided into three general
cases :

1°. When the work after completion coincides with
the prescribed lines, the material remaining to be removed is nothing, and the preliminary calculation serves
for the final.

2°. When the greater portion of the work coincides
with the proper slope and base, but a few sections are
found to be defective or redundant, the vols., representing the work remaining to be done, bet. the latter or
affected by them should be calculated, and the result
subtracted from the result of the whole original calculation. To illustrate this, suppose the finished work of
the example of excavation, last given in connection with
preliminary estimates, to coincide with the prescribed
lines at every cross-section except the first two. Let
the final notes of these, together with those of the third
cross-section, used to calculate second vol., be the following:

Sta.	Centre	Left	Right	Remarks
1	0	$\frac{2}{16}, \frac{2}{17}, \frac{2}{18}$	$\frac{2}{18}, \frac{3}{18.5}$	
+50	−1	$\frac{1}{15.5}, \frac{-1}{15}$	$\frac{-1}{18}, \frac{3}{18.5}$	18.5 |
2	0	$\frac{0}{18.5}, \frac{0}{15}$	$\frac{0}{15}, \frac{4}{17}$	

At sta. 1, as shown in diagram, the base has been cut too wide on the right, and the left slope flattened too

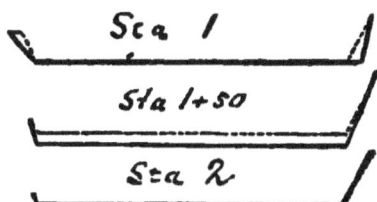

much. At sta. 1+50 the work has been cut too deep. The 2d break on left of sta. 1 runs to left slope stake of sta. 1+50 and is so recorded in REMARKS.

The sections being equi-distant, both vols. may be calculated together. Half the widths and prods. of first and last sections are used, according to rule. There being no interms., the 2d col. of prods. need not be constructed. The pris. cor. is inconsiderable. The result is −59.7. Subtracting this from the result of original calculation, the true content excavated is left.

TABLE OF OPERATIONS.

STA.	WIDTHS.		CEN	PRODS.	CORRECTIONS. INTERMS PER.	BREAKS.
1	8.	16.	2	16.		15.5 × 1 ⎤
"	8.25	16.5	0	0		⎥
"	L	15.	−2	−15.		15.0 × −2 ⎥
"	L	17.	−2	−17.		15.5 × −2 ⎦
"	R	16.	−3	−24.		
+50	15.5	15.5	−1	−15.5		
"	16.5	16.5	−1	−16.5		
"	L	15.	−2	−30.		
"	R	15.	−4	−60.		
2	7.7	15.4	0	0		
"	8.5	17.	0	0		
"	L	15.	−0.8	− 6.		
"	R	15.	−4.	−30.		

$$64.45 \qquad\qquad -198. \qquad 0 \quad 0$$
$$Sbw = \qquad +1933.5$$
$$-2Sl^2n = \qquad -1800.$$
$$\overline{2)\,-64.5}$$
$$\overline{9)\,-3225.}$$
$$\overline{6)\,-358.3}$$
$$\overline{-(-59.7)}$$
$$5083.27$$
$$\overline{5143.} = no.\ of\ cu.\ yds\ excavated.$$

3°. When a large number of the sections do not coincide with base and slope lines, or when such irreg. sections are scattered and isolated, making many separate calculations necessary, it is better to make one operation of the final estimate, considering every section in the work, whether its area be zero or not. There is one important advantage available in this case. Thus, the formula for original approx. content is

$$\left\{ \begin{array}{c} mid\ prods. + \tfrac{1}{2}\ end\ prods \\ + Sb(mid\text{-}widths + \tfrac{1}{2}\ end\text{-}widths) - 2Sl^2n \end{array} \right\} \frac{D}{54},$$

and the formula for approx. final content is the same; wherefore, since the widths are identical in both, the terms

$$Sb(mid\text{-}widths + \tfrac{1}{2}\ end\text{-}widths) - 2Sl^2n$$

vanish by subtraction. Hence, in the final calculation, the 1st col. of widths need not be constructed, summed and multipled by Sb, and the quantity $-2Sb^2n$ need not be added, but the col. of prods. only considered, to this added the cor. for interms. and the pris. cor., and the sum subtracted from the sum of prods., the cor. for interms. and the pris. cor. of the original calculation.

Between the cross-sections, which do not coincide with slope, there are in this case to be considered sections which do coincide, and whose areas are there-

fore zero. These are easily disposed of. In common with the other sections their widths $\times Sb$ need not be considered. Their prods. are for each

$$\tfrac{1}{2}vm + \tfrac{1}{2}xn + \tfrac{1}{2}b(c-l) + \tfrac{1}{2}b(c-r).$$

The first two terms are always zero, and the last two are always $-\tfrac{1}{2}b(r+l)$. Since the factor b is common to all these sections, the prods. of all are found at once by adding all the side-heights, prefixing the minus sign, and multiplying by b, following the rule, however, of using the half-prods. of sections at full stations preceding interms., and the half prods. of last interms. So, add to the side-heights of all not followed by interms. the half-heights of those followed by interms. and the half-heights of last interms., before multiplying by b. This is all the work required for these sections, since the cor. for interms. and the pris. cor. always reduce to zero. The former cor.,

$$\left(P - P' + Sb(w-w')\right)\tfrac{D}{108}.$$

where P, P', are the respective prods., reduces to

$$\left(-b(r+l-r'-l') + Sb(w-w')\right)\tfrac{D}{108}.$$

But $\qquad w = \tfrac{r}{s} + \tfrac{l}{s} + 2b, \quad w' = \tfrac{r'}{s} + \tfrac{l'}{s} + 2b.$

$\therefore \qquad Sb(w-w') = b(r+l-r'-l'),$

and the whole cor. reduces to zero.   The pris. cor. is

$$\left\{ \begin{array}{c} (v-v')(m'-m)+(x-x')(n'-n) \\ +(b-b)((c'-l')-(c-l))+(b-b)((c'-r')-(c-r)) \end{array} \right\} \frac{D}{324},$$

and each term reduces to zero under the conditions.

Let us suppose that the first two sections of finished work of the excavation, last treated as an example of preliminary calculation, are as noted in second case, and that the last four sections also of the same excavation do not coincide with the true slope. It now becomes more profitable to estimate the material, remaining to be excavated, in one operation. The final notes of the cut are the following.

FIELD NOTES.

STA.	CENTRE.	LEFT.	RIGHT.	REMARKS.
1	0	$\frac{2}{18},\frac{2}{17},\frac{0}{18}$	$\frac{0}{18},\frac{3}{18.5}$	
+50	−1	$\frac{1}{13.5},\frac{-1}{14}$	$\frac{-1}{13},\frac{3}{18.5}$	
2	0	$\frac{0}{13.4},\frac{8}{14},\frac{12}{14}$	$\frac{0}{15},\frac{4}{17}$	16.5
3	0	$\frac{0}{15}$	$\frac{0}{15},\frac{8}{19}$	
+20	0	$\frac{0}{15}$	$\frac{0}{15},\frac{14}{22}$	
4	0	$\frac{0}{15}$	$\frac{0}{15},\frac{16}{23}$	
5	0	$\frac{4}{17},\frac{0}{15}$	$\frac{0}{15},\frac{18}{24}$	
6	0	$\frac{0}{15}$	$\frac{0}{15},\frac{23}{18.5},\frac{24}{27}$	
7	0	$\frac{0}{15}$	$\frac{0}{15},\frac{22}{23},\frac{30}{30}$	
8	0	$\frac{0}{15}$	$\frac{0}{15},\frac{34}{34},\frac{28}{28}$	
9	0	$\frac{0}{15}$	$\frac{0}{15},\frac{30}{28}$	26

Table of Operations.

STA.	WIDTHS.		CEN.	PRODUCTS.	CORRECTIONS. INTERMS.	PRISMOIDAL.	BREAKS.	
1		16..	2	32.			15.5 × 1	sta. 1+50 for Sta. 1.
		16.5	0	0			15 × − 2	
	L	15.	− 2	−30.			15.5 × − 2	
	L	17.	− 2	−34.				
	R	16.	− 3	−48.				
+50		15.5	− 1	− 7.75	−250	0	15.4 × 0	sta. 2 for Sta. 1+50.
		16.5	− 1	− 8.25			17 × 0	
	L	15.	− 2	− 15.			15 × − 0.8	
	R	15.	− 4	− 30.			15 × − 4.	
2....5		15.	−53.8	−807.		0		
6		15.	0	0	0		17 × 0	sta. 6 for Sta. 6.
		27.	23	+621.	−15		24 × 18	
	R	15.	−23	−345.	0		15 × − 4	
	R	25.5	−24	−612.	+2/0	−600	15 × −18	
7		15.	0	0	+3		24 × −18	
		30.	22	+660.				
	R	15.	−22	−330.	0			
	R	25.	−30	−750.	−3/0	0		
8		15.	0	0	−2		25 × 20	sta. 7 for Sta. 8.
		29.	20	+580.	0		16 × −20	
	R	15.	−20	−300.	+2/0	0	25 × −20	
	R	24.	−28	−672.				
9		15.	0	0	0			
		25.	0	0	0			
	R	16.	−20	−160.	−8	−800		

$$-(-2176.) \qquad -(-250) \quad -(-1400)$$
$$565.2 \qquad -2640 \qquad -14420$$
$$\overline{2741.2} \qquad \overline{-2390} \qquad \overline{-13020}$$

Cor. for Int's ÷ 200 = − 11.95
Pris. Cor. ÷ 600 = − 21.7

9)270755.
6)30083.9

5014. cu. yds.

The calculation of this remaining work is precisely similar to the original calculation, except the omission of the 1st col. of widths. The sections of stas. 2, 3, 3+20, 4, 5, coinciding with slope lines and lying adjacent, are noted together, and their prods. joined in one. At sta. 2 the sum of side-heights is 4. 8. The sum of side-heights at sta. 3 is 8 : half should be taken since it precedes an interm. Half of 14 should be taken; because it is the sum of heights of last interm. The whole heights at stas. 4 and 5 should be taken. The sum is 53.8, to which prefix minus sign. The common multiplier is 15. Since there is but one interm. to cause a cor., the 2d col. of prods. has not been constructed further than the first section. For this interm., subtract width of sta. 2 from that of sta. 1, and multiply dif. by 30 [$=Sb$]. The result is 3. Subtract prods. of sta. 2 from those of sta. 1. The dif is -8. $(3-8)50 = -250$. The pris. cor. for the first two vols. is inconsiderable, as discovered in 2d case. The cor. bet. sta. 2 and sta. 5 is nothing, because the work bet. them everywhere coincides with the true slope and base. The right of sta. 5 is arranged in col. of remarks to correspond with right of sta. 6. The right of sta. 9 is arranged to correspond with that of sta. 8. Subtract the results of the three cols. from the results of the three corresponding cols. of the original calculation. Use the same factors with the differences that were used with results of same columns in original calculations. The result is the exact amount of the mass of material removed.

This can be conveniently divided into the classes of material, as rock, slate, common earth, etc., from the monthly estimates or from the final notes, exactly as it is now done after obtaining the whole amt. by the ordinary method. A fair plan is to use the proportion : as the amt. of one class of work, as rock, in the sum of monthly estimates, is to the whole amt. of material in the same, so is the true amt. of rock to the true total amt. of material, as just found, unless extra trimming of earth or other material has been done between the two measurements. But these details are well understood, and form no part of the plan of this book.

CHANGE OF SLOPE.

It is often found necessary after the preliminary calculation to make the slope flatter than at first intended,
or advisable to make it steeper, according to the material struck. If the surface of ground be regular, as shown
in the first diagram, it is better to calculate the whole
work anew with the new ratio of slope and the

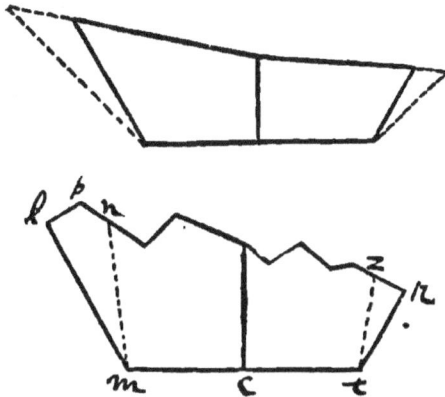

new widths. But, if the surface of ground be very irreg.,
as shown in next diagram, it becomes preferable to avoid
the recalculation of the irreg. part, by considering the
new slopes as the under part of perimeter of section of
excavation, and calculating the work remaining to be
done, to reach the original slopes, as in the example last
given of finished work, and finally subtracting the latter
result from the result of original calculation. The
formula for the cross-section of remaining work in last
diagram is

$$\tfrac{1}{2}v(p+Sb)+\tfrac{1}{2}x(z+Sb)+\tfrac{1}{2}b(c-n)+\tfrac{1}{2}n'(m-p)$$
$$+\tfrac{1}{2}p'(n-')+\tfrac{1}{2}b(c-z)+\tfrac{1}{2}z'(t-r).$$

In this instance it represents a positive quantity, because
the new slopes are steeper than the old : if the new ones
were flatter, the section remaining would be negative.

5

RUDE PRELIMINARY ESTIMATES.

For a rough, early estimate, when the widths of the sections are known, the first rule for calculating a series of equi-lengthed vols. is recommended, disregarding interms., breaks and the pris. cor., and taking the cross-sections as far apart as possible. This computation can be made very rapidly.

When the widths have not been calculated or staked out, and the preliminary notes only are available, these consisting of a record of centre elevation and the elevation of a point on each side of centre 50 ft. distant, at each sta., a rough but hasty estimate can be made, by placing centre heights above grade in one column, and the squares of these in another, conveniently extracted from a *table of squares* such as Henck has furnished in his FIELD BOOK, and treating their sums as in the treatment of level sections, already discussed, disregarding altogether the third column, constructed for the pris. cor.

When the surface line at right angles to centre line is unbroken, as sometimes occurs at a great many consecutive stas., the areas of the sections, under supposition that they are level and of the same height as centre, are always too small, as seen in the diagram, by the triangle

$m\ r\ r'$, and, consequently, the resulting mass of material between is also smaller than the true contents. This fact has often been pointed out, but we have seen no formula of cor. for the error. Such an one, that serves very well both for accuracy and brevity, can be obtained in the following way. $m\ r\ r'$ is similar to $l\ e\ r$. $eg \div b = S : \therefore eg = Sb$. Let A' represent area of $l\ e\ r$, and let the abbreviated word *Cor.* denote the area of $m\ r\ r'$. Then

$$Cor : A' :: z^2 : (c + Sb)^2.$$

Let A denote the area of $l'\ e\ r'$, equal to $A' - Cor$. Now by the principles of proportion

$$Cor : A' - Cor :: z^2 : (c + Sb)^2 - z^2, \text{ or } Cor = \frac{z^2}{(c+Sb)^2 - z^2} A.$$

If h be the elevation at 50 ft. from centre on higher side, then

$$z : h - c :: b + \tfrac{c}{S} : 50, \text{ or } z = \tfrac{1}{50}(h-c)(c + Sb)\tfrac{1}{S}.$$

$$Cor = \frac{(\tfrac{1}{50})^2(h-c)^2(c+Sb)^2(\tfrac{1}{S})^2}{(c+Sb)^2 - (\tfrac{1}{50})^2(h-c)^2(c+Sb)^2(\tfrac{1}{S})^2}A \qquad = \frac{(\tfrac{h-c}{50})^2}{S^2 - (\tfrac{h-c}{50})^2}A.$$

It is here seen that the ratio of the correcting area to A is as $\dfrac{(\tfrac{h-c}{50})^2}{S^2 - (\tfrac{h-c}{50})^2}$ is to unity; and, if these areas be moved through any dist. D, the vols. generated are in the same ratio. Therefore, if the average area of $m\ r\ r'$ in all the sections through the work be ascertained and multiplied by D, just as the average of $l'\ e\ r'$ is multiplied by D, the resulting vols. are related by the ratio $\dfrac{(\tfrac{h-c}{50})^2}{S^2 - (\tfrac{h-c}{50})^2} : 1$, where h is the average higher side-height, and c the average centre-height. But in calculating the vol. generated by moving the average sectional area $l'\ p\ k\ r'$ through the dist. D, a column of centres has been constructed and summed. Let C be this sum. Then $\tfrac{C}{n}$, where n is the no. of vols., is the average centre-height. Similarly, if H be the sum of side-heights, the higher one only being taken at each sta., or either if the section be level, then $\tfrac{H}{n}$ is the average side-height. So the vols. hold the ratio

$$\frac{\left(\frac{H-C}{50m}\right)^2}{S^2-\left(\frac{H-C}{50m}\right)^2}:1 \qquad \text{or} \qquad \frac{\left(\frac{H-C}{50}\right)^2}{S^2 n^2-\left(\frac{H-C}{50}\right)^2}:1.$$

If, therefore, after finding the volume generated by moving the average area $l'pkr'$ through D, and adding thereto the volume beneath road-bed, Sb^2D, the sum be multiplied by $\frac{\left(\frac{H-C}{50}\right)^2}{S^2 n^2-\left(\frac{H-C}{50}\right)^2}$, the prod. is the cor. to be added to the original volume.

This shall be illustrated from the following notes. S is considered $\frac{1}{4}$, b, 10 ft.

STA.	LEFT.	CENTRE.	RIGHT.
1	1	6	11
2	2	8	14
3	−1	9	19
4	−5	15	35
5	15	20	25
6	18	14	10
7	12	10	8
8	11	8	5
9	12	7	2

STA.	c	c^2	h
1	3	18	0.5
2	8	64	2.
3	9	81	−1.
4	15	225	−5.
5	20	400	15
6	14	196	10
7	10	100	8
8	8	64	5
9	3.5	24.5	1
	90.5	1172.5	35.5
		1810.	50)55.0
		2982.5	1.1
		800.	62.79)1.21(.0193
		3782.5	6279
		.0193	5821
		1.135	56511
		34.042	1699
		37.825	
		9)305550.2	
		3)33950.02	
		11316.67 cu. yds.	

The cols. headed c, c^2, are constructed as for level-sections, the sum of former multiplied by 2 b, the sum of latter by $\frac{1}{3}$, and the prods. added together. Next a col. of heights is made, using half the first and last as with centres. Instead of taking the higher side elevations, the lower ones have here been used, and their sum subtracted from sum of centres: the dif. is the same and the numbers to be dealt with smaller. This dif. divide by 50: square the quotient for a dividend: for a divisor subtract dividend from $S^2 n^2$, 64 in this example. The quotient multiplied by the whole volume down to where the slopes meet, is the cor. to be added to original volume. Therefore, to 2982.5 add $S^2 n$ [=800], and multiply sum by .0193. The three partial prods. form the cor., which, added to the 2982.5, yields a sum to be multiplied by 100 and divided by 27 to produce a close approximation to the number of cu. yds.

The peculiar merit of this plan, as applied to preliminary estimates, is the facility with which changes in the result may be made to correspond with alteration in slope, width of base, altitude of grade, etc. For instance, to alter the result for a change of slope, the cols. need not be touched, but simply the values underneath modified by the new value of S instead of the old, where S enters the calculation. Similarly a change may be effected for a change of width in base. If it be desired to ascertain the effect of sinking the grade all the way through, say 10 ft., merely add to sum of centres 10 n [=80], to the sum of squares $\overline{10}^2 n$ [=800] plus $2 \times 10 \times$ sum of centres [=1810]. The cor. ratio, .0193:1, need not be altered. Therefore, the increase of volume equals

$$4210 \times 1.0193 \times 100 \div 27.$$

The general formula for this example is, letting x represent the proposed increased cutting and M the resulting increase in the content to be excavated,

$$[nx.2b + (nx^2 + 2xC)\tfrac{1}{3}]\tfrac{101.93}{27} = M,$$

C being the sum of col. headed c. Solving with respect to x, we have

$$x = \tfrac{1}{n}\left[\sqrt{\tfrac{27.3n}{101.93}M + (C + Sbn)^2} - (C + Sbn)\right].$$

Suppose we want 1000 cu. yds. more from the cutting. Let $M=1000$, when $x=+0.7\frac{1}{2}$ ft. To make the excavation 1000 cu. yds. less, let $M=-1000$, when $x=-0.79$ ft. For any other example simply use instead of 101.93 the proper factor, or, if the cor. is not used, take D alone, generally 100. If the centre line be moved to a position on the right or left, parallel to its present position, the new centres must be tabulated and squared, but the ratio, .0193 : 1, is still correct.

It is seen that the cor. is not quite 3 per cent. of the true contents in this example. If such close work be not required, the cor. need not be noticed. But, if it be retained, the result is the exact approx. contents of the vols. bet. the actual cross-sections, afterward to be staked out, when the pris. cor. may be made and attached to the preserved result.

For a hasty estimate, when the surface line is broken at centre, allowance must be made, in the centre height used, for the direction of the angle's concavity. If this be downward, the centre used for equal level-section must be diminished, and, if upward, increased. Repeated trial, tested by computation of the true areas, gives skill in this. The allowance varies with the rate of slope, width of base, centre-height, and the size of the angle's concavity. But, since the base and slope remain constant through the work, the centre-height and concavity only exert influence in a regular cutting. The cor. need not be applied in this case. Without this latter rather rough method recourse must be had to much longer rules.

CORRECTION FOR

EXCAVATION ON CURVES.

THE error of calculating earthwork on a curve, disregarding the curvature, is not great. The manner of calculating exactly the content of a vol. is the following. The diagram here accompanying represents a vol. on a curve of exaggerated degree of curvature. $c\,H$ is the radius, $= R$, H the angle subtended by the distance. $c\,c'$,

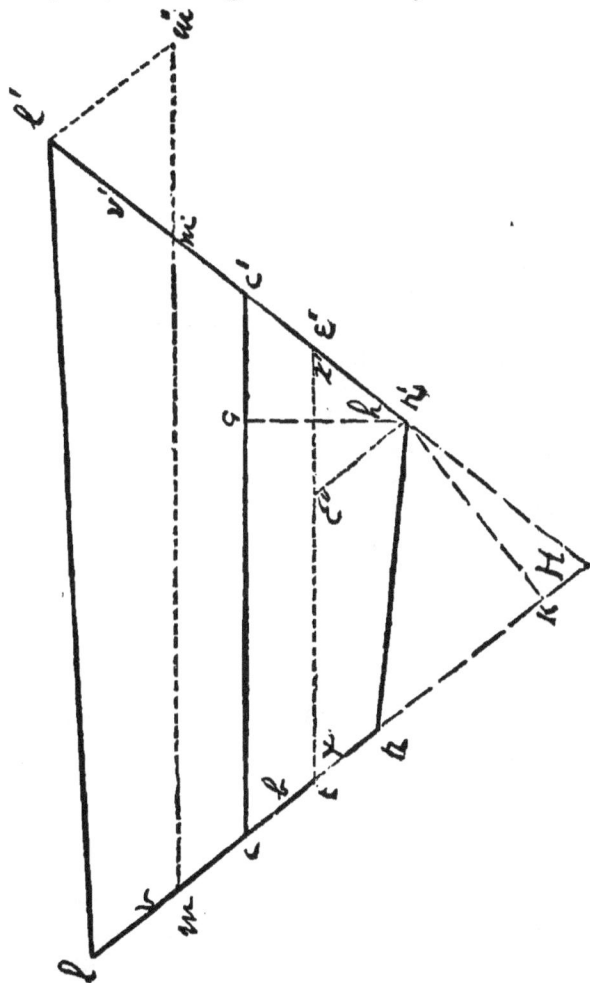

$=D$, between two stations. h is the angle of deflexion. $e\,e'$, $m\,m'$ are the edges of road-bed. Conceive a straight line joining c and r'. Estimate the vol. $c\,c\,r'$ by multiplying the area of its base, $\frac{1}{2}cc' \times qr', =\frac{1}{2}Dx'cos h$, by the average height, $\frac{1}{3}(c+c'+r')$. Add to this the vol. $c\,r\,r'$, whose base is $\frac{1}{2}x \times kr', =\frac{1}{2}x(R-x')sin H$, and average altitude $\frac{1}{3}(c+r+r')$. If now the diagonal vertical plane $r\,c'$ be considered, and the vols. $c\,c'\,r, c'\,r'\,r$ be similarly calculated, and their sum averaged with the sum of the two other vols., that average is the exact content bet. the warped surface $c\,r\,r'\,c'$, the plane of grade and vertical planes through the perimeter of warped surface. This principle was also used on page 54, in connection with end-volumes. It is easy of demonstration. A proof of it is given on page 379 of Gillespie's ROADS AND RAILROADS. It is also found in SONNET'S DICTIONNAIRE DES MATHÉMATIQUES APPLIQUÉES, and in many other works. Subtract from this vol. the triangular prism $e\,r\,r'\,e''$, whose end-section $r'\,e''$ is exactly equal and similar to $r'\,e'$ and whose altitude is $k\,r'$; also the pyramid $e''\,r'\,e'$ of altitude r' and basal area $\frac{1}{2}(x'-b)^2 sin H$. In the same manner compute the vol. $l\,c\,c'\,l'$, bounded by vertical planes, subtract the prism $l\,m\,m''\,l'$, and add the pyramid $l'\,m'\,m''$. This process is entirely too laborious for any practical application except to test the accuracy of available methods.

John B. Henck has instituted a formula of cor., which is very acceptable on account of its simplicity and the accuracy of its results in all ordinary cases. Expressed in his own symbols, d, d', being side-widths, h, h', side-heigths, this formula is
$$\left[\tfrac{1}{2}c(d-d')+\tfrac{1}{4}B(h-h')\right]\tfrac{D(d+d')}{3R}.$$
Instead of $\frac{D}{R}$ may be used $2sin h$. This, translated, becomes in cu. yds.,

☞ $$w(v-x)(c+Sb)\tfrac{sin h}{3T}.$$

The dif., $v-x$, is found by subtracting the inner side-width at a sta. from the outer. The cor. is thus positive when the outer side is greater, and negative when it is the lesser. The cor. is applied at each sta., half the results being accepted at the end sta. of a curve.

The following table shows the amts., resulting from the use of Henck's formula, compared with the true contents, ascertained by the lengthy method explained at head of this article, of four separate vols. on a 4° curve. The 1st col. contains the true amts., obtained by prismoidal formula, without regard to curvature. Half of Henck's cor. is used at each sta., as this is supposed to be the end of the curve. Prefixed to the table are the field-notes of the cross-sections used. $S=\frac{1}{4}, b=10, D=100, R=1432\ 69, sinh=.0349, cosh=.99985, sinH=.06976.$

	CENTRE.	LEFT.	RIGHT.
1st Vol.	10	$\frac{30}{0}$	$\frac{40}{0}$
	30	$\frac{40}{0}$	$\frac{40}{0}$
2nd Vol.	10	$\frac{40}{0}$	$\frac{30}{0}$
	30	$\frac{40}{0}$	$\frac{40}{0}$
3rd Vol.	10	$\frac{30}{0}$	$\frac{5}{15}$
	10	$\frac{30}{0}$	$\frac{5}{15}$
4th Vol.	10	$\frac{10}{0}$	$\frac{10}{0}$
	10	$\frac{10}{20}$	$\frac{10}{20}$

	TRUE AMT WITHOUT CURVATU'E	HENCK'S COR.	CORRECT'D AMT.	TRUE AMT.	ERROR.
1st Vol	128333.33	+174.5	128507.83	128239.60	+268.23
2d Vol	128333.33	+337.36	128670.69	128623.27	+47.42
3d Vol	35000.	+157.75	35157.75	35151.9	+5.85
4th Vol	30000.	0	30000.	29995.5	+4.5

The surface of 1st vol. is very much warped, the outside of one section being the lower, and the outer of the other the higher, side; and it is seen that Henck's cor. is here much at fault, the result of the calculation without noticing the curvature being much nearer the true content than is the corrected amount. The next vol. is the same as last, except that its first section is reversed so that the higher sides are both outer. The surface of ground is, therefore, very much less warped, and, as seen, the cor. of Henck is very much nearer the truth. The next vol. is bounded by cross-sections of same size and shape. The surface is, therefore, two planes, and Henck's cor. errs but little. The 4th vol. is level. The inference is: *Where the surface of ground is a plane or nearly so, Henck's formula is very correct.* A much-warped surface is rare; and, when it does occur, it is, perhaps, better not to attempt a correction.

BORROW PITS.

For the calculation of any class of cross-sections, whether there be a regular slope and base or not, we recommend the method, employed here for intermediate sections and for breaks, of using the distance of each point from the base line as a factor with the dif. of elevation of the two adjacent points. The object is to secure one col. of very small factors, the differences, although the other col. receives larger factors than by the ordinary method; also to facilitate the application of the pris. cor. where required. Let the following be the field notes of a borrow-pit, whose datum plane is fixed at 20 ft. below the elevation of sta. 0 on base-line.

STA.						
0	$\frac{20}{0}$	$\frac{16}{0}$	$\frac{18}{0}$	$\frac{18}{0}$	$\frac{24}{0}$	$\frac{20}{100}$
1	$\frac{15}{0}$	$\frac{12}{0}$	$\frac{18}{0}$	$\frac{14}{0}$	$\frac{28}{0}$	$\frac{20}{100}$
2	$\frac{10}{0}$	$\frac{12}{0}$	$\frac{14}{0}$	$\frac{20}{0}$		$\frac{24}{100}$
3	$\frac{15}{0}$	$\frac{18}{0}$	$\frac{28}{0}$		$\frac{24}{0}$	$\frac{24}{100}$
4	$\frac{20}{0}$	$\frac{22}{0}$	$\frac{24}{0}$		$\frac{30}{0}$	$\frac{27}{100}$

To estimate the mass of earth bet. this surface and datum plane, arrange distances in one col., as shown, and the differences of adjacent elevations in the next, except for the last distance, 100, opposite which place the sum of last two elevations. From these calculate the col. of prods., using half the prods. of first and last stas. Multiply the sum by the distance bet. stas., 25 ft., and divide by 54.

It is scarcely necessary to mention the points of brevity in favor of this plan, as against the ordinary method of calculating such work, since this topic has been fully discussed already; but we might recall the fact that

TABLE OF OPERATIONS.

STA.	DIST.	$h-h'$	PRODS.	CORRECTIONS. INTRMS. PRISMOIDAL.
0	10	10	50.	
	30	0	0	
	50	−15	− 375.	
	75	− 5	− 187.5	
	100	45	2250.	25
1	15	5	75.	0
	30	− 5	− 150.	25
	45	−10	− 450.	0
	70	− 5	− 350.	0
	100	40	4000.	0
2	15	− 2	− 30.	50
	40	−10	− 400.	− 30
	60	−13	− 780.	0
	100	45	4500.	0
3	20	− 5	− 100.	15
	35	− 8	− 280.	10
	75	− 5	− 375.	−120
	100	51	5100.	0
4	20	− 7	− 70.	0
	30	− 8	− 120.	0
	80	0	0	− 25
	100	57	2850.	0

For Sta. 1. $\{\begin{smallmatrix}60\times-8\\60\times-5\end{smallmatrix}\}$

$$15157.5 \qquad\qquad -50$$

$$\textit{Pris. Cor.} \div 6 = \quad -8.33$$
$$4\overline{)15149.17}$$
$$9\overline{)378729.25}$$
$$6\overline{)42081.03}$$
$$7013.5 \text{ cu. yds.}$$

commonly the areas of the several parts of each cross-section are combined to find the area of that section, the latter value being used, while here all these partial areas are added at once. Also the factors used are more readily handled, since those of the col. of differences are nearly all composed of a single digit only, making unnecessary the extra work of a separate multiplication, while the larger ones opposite the last distances need for the operation merely a shifting of decimal point. Besides this, the measurements are now arranged in perfect order for the easy application of the pris. cor. Simply multiply the dif. of corresponding distances by the inverse dif. of corresponding factors in col. of differences, and multiply by dist. bet. sections for a value to set in col. of cor. The last pair of factors at each sta. thus need no cor. in this example, but would require it, were the far side of the pit irregular. Sta. 2 has one point less recorded than sta. 1. The last break of latter is noticed to merge into last of former; therefore, in applying cor. the last break of sta. 2 is considered to be two breaks, and the new factors are recorded at the side of the old ones in the table, to be used only with the section before. Since the vols. are all of one length, divide cor. by 6 and add to sum of prods. Multiply sum by 25, $=\frac{100}{4}$, and divide by 54. If there were interm. cross-sections, use the half prods. of stas. next before and of last interms., as in road-way calculation, and for the cor. multiply the dist. of each interm. from sta. next before by the dif. of prods. of sections adjacent. Divide sum of col. by 108.

By this plan, as illustrated, it is seen that the pris. cor. is very easy of application, and that it is also readily discerned where the cor. is and where it is not required. For instance, where a knoll or the spur of a hill has been removed, the approximation by end areas is extremely faulty, and the cor. assumes great importance.

In concluding, we must again say that the principles, upon which this plan of computing earthwork is founded, are few and simple, and the results, excepting the rough estimates, perfectly accurate; that the calculator soon becomes familiar with the processes, since the values naturally find their proper places in the tables of operation; and that these tables themselves are neat and regular, recording all the work as the ever-present authority for the results they produce, and always ready for a rapid review, while by ordinary methods the whole work is thrown away, and nothing preserved but the areas of the cross-sections or the contents of the vols. between.

GENERAL NOTE.

VOLUMES BOUNDED BY WARPED SURFACES.

IN staking out earthwork, the sections may be taken as far apart as the ground continues to change its slope uniformly, no matter how much the surface may warp, nor how great the distance between cross-sections; that

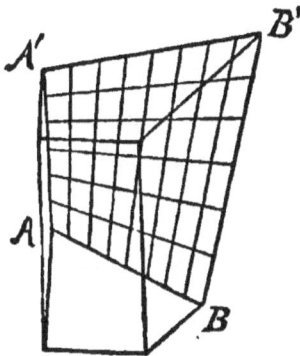

is, while the lines AA' and BB' remain straight, and the surface from AA' slopes to BB' in straight lines. The most rational conception of the surface bet: four straight bounding lines is that it is the *hyperbolic-paraboloid* or *warped surface*, which may be generated by moving AB, parallel to the planes of end sections, along AA', BB', till it assumes the position $A'B'$, or by moving AA' along AB, $A'B'$ till it becomes BB', each end traveling with speed proportional to the length of its path. To the vol. beneath such a surface the prismoidal formula applies exactly. The assumption by Henck of one straight diagonal from A' to B or A to B', has received severe criticism, and it, moreover, has not the advantage of making tho formulæ less intricate.

Little fault, however, can be found with the field work of engineers, since the rule of placing cross-sections so that they shall be connected by straight lines is carefully regarded in preparing notes for any mode of calculation, the lack of straightness in a longitudinal element showing the presence of an intermediate irreg. section.

The principle that the prismoidal formula applies to vols. beneath warped surfaces was assumed on page 21, since it has been so often used and shown before. A general proof of this application to vols. bet. three-level sections, whether covered by plane or warped surfaces, can be deduced by calculus, as Prof. Gillespie has done for a vol. bet. two trapezoids in his MANUAL OF ROAD-MAKING. Let $\frac{1}{2}w(c+Sb)-Sb^2$, $\frac{1}{2}w'(c'+Sb)-Sb^2$ be the areas of end sections. At a distance x from the first the centre of a cross-section is $c+(c'-c)\frac{x}{D}$, width $w+(w'-w)\frac{x}{D}$. Hence its area is

$$\tfrac{1}{2}\big(w+(w'-w)\tfrac{x}{D}\big)\big(c+(c'-c)\tfrac{x}{D}+Sb\big)-Sb^2.$$

Multiplying this by dx, after performing the multiplication indicated, we obtain the differential of the vol.,

$$\tfrac{1}{2}\left\{\begin{array}{l} w(c+Sb)dx+(w'c'-w'c-wc'+wc)\dfrac{x^2dx}{D^2} \\[2mm] +\big(wc'-2wc+w'c+Sb(w'-w)\big)\dfrac{xdx}{D} \end{array}\right\} - Sb^2dx.$$

Integrating bet. the limits 0 and D, we obtain

$$Vol=\big(w'c+wc'+2(w'c'+wc)+3Sb(w+w')-12Sb^2\big)\tfrac{D}{12},$$

which is identical with the formula for the true vol. on page 21. Irreg. vols. may be divided by vertical planes through corresponding breaks into portions, which may be shown in a manner similar to the above to be subject to the prismoidal formula : therefore the composite vols. are subject to the prismoidal formula.

On page 40 it is advised to note the vanishing extremities of breaks, and subsequently the importance of this is illustrated by examples. The general discussion of this topic can be readily effected by the aid of calculus in the manner following. The diagram represents a triangular, warped-faced wedge. Its base is ABC, edge DE: the back of the wedge is the face $ADEC$, warped or plane according to the relative positions of the lines AC and DE. The remaining faces, divided by the edge BH, warp from AB to DH and BC to HE. It will be noticed that this wedge is similar to

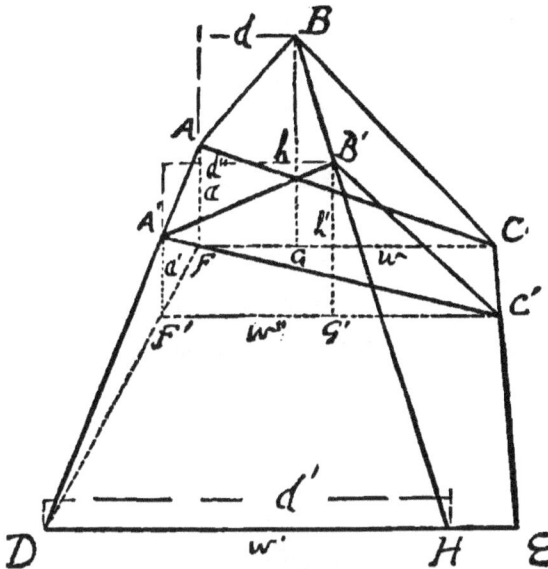

the wedge $ABCDE$ shown in the diagram of vol. on page 41, and that all the irregularities of rail-road vols., which vanish at one end of the vol., can be decomposed into such wedges as the one here shown. In the discussion it is convenient to make the edge DE horizontal. Using the symbols of the diagram for measurements, we have for a section distant from the base by x

$$a' = \frac{a(D-x)}{D}, w'' = w + (w'-w)\frac{x}{D}, \quad d'' = d + (d'-d)\frac{x}{D}, h' = \frac{h(D-x)}{D} \; ;$$

and for the area of this section

$$A'B'C' = A'B'G'F' + B'C'G' - A'C'F' =$$

$$\tfrac{1}{2}d''(a'+h') + \tfrac{1}{2}h'(w''-d'') - \tfrac{1}{2}a'w'' = \tfrac{1}{2}(a'd'' + h'w'' - a'w'') =$$

$$\tfrac{1}{2} \left\{ \begin{array}{l} \dfrac{ad(D-x)}{D} + \dfrac{a(d'-d)(Dx-x^2)}{D^2} \\[2mm] (h-a)\left(\dfrac{w(D-x)}{D} + \dfrac{(w'-w)(Dx-x^2)}{D^2} \right) \end{array} \right\} \cdot$$

Multiplying by dx we obtain the differential of the vol. The indefinite integral is

$$\frac{1}{4}\left\{ \begin{array}{c} \dfrac{ad(Dx-\frac{1}{2}x^2)}{D}+\dfrac{a(d'-d)(\frac{1}{2}Dx^2-\frac{1}{3}x^3)}{D^2} \\ (h-a)\!\left(\dfrac{w(Dx-\frac{1}{2}x^2)}{D}+\dfrac{(w'-w)(\frac{1}{2}Dx^2-\frac{1}{3}x^3)}{D^2}\right) \end{array} \right\}.$$

Limiting this by 0 and D, we have

$$Vol. = \tfrac{1}{4}\Big(\tfrac{1}{2}adD+\tfrac{1}{6}a(d'-d)D+(h-a)(\tfrac{1}{2}wD+\tfrac{1}{6}(w'-w)D)\Big)=$$

$$(a(2d+d')+(h-a)(2w+w'))\tfrac{D}{12}.$$

By this formula it is instantly seen that to give d' any increment m increases the vol. by

$$\frac{amD}{12},$$

while to give d the same increment increases the vol. by

$$\frac{2amD}{12}.$$

Therefore, *to move the vanishing end of a ridge or hollow any distance alters the content of a vol. exactly half as much as to move the other end of the same ridge or hollow an equal distance in the same direction, as stated on page* 43. It is also to be remarked that when $a=0$, d and d' vanish from the formula. Therefore, *When DE is parallel to AC, or when the back of the wedge is a plane surface, to move the vanishing end of a ridge or hollow any distance does not alter the content of the vol.: but the further AC, DE are from being parallel, or the more warped is the back of the wedge, the greater is the effect, of moving the vanishing end of a ridge or hollow, upon the content of a vol., as stated on page* 44.

We consider this an important matter since we have found no book which demonstrates the consequence of fixing these fading ends. That it has a consequence is shown by the example on page 41. The irregularity of one section in this example is made very marked in order to make visible, if possible, the increase of content occasioned by moving m, n, p to M, N, P. But, if the irregularity were scarcely perceptible, as in the most ordinary example, the increment $\frac{amD}{12}$ would be the same, since it depends entirely and only upon the difference of inclination of AC, DE, as represented in the formula by the symbol a.

Prof. Gillespie proposes in an off-hand way [ROADS AND RAILROADS, page 373], considering such a vol., to conceive vertical planes passed through the breaks of irreg. section and cutting the surface line of the other section proportionately, preparatory to the application of prismoidal formula. But this assumption that the ridges and hollows cut both sections proportionally has the same fault that Prof. Gillespie finds in Henck's diagonals, that they would not always happen so. For, supposing them to be so now, if hereafter, proceeding from one section, the direction of the centre line be changed, both the ridges of Henck and of Gillespie must change also or break the law. There is no more reason in assuming that the ridges vanish at these points than at points fixed by any other arbitrary rule. One thing is remarkable about Prof. Gillespie's method. If the approx. content of the wedge be subtracted from the true, the difference is

$$(a(d'-d)+(h-a)(w'-w))\tfrac{D}{12}.$$

If $w=w'$, the difference is $a(d'-d)\tfrac{D}{12}$. But on Prof. Gillespie's assumption d and d' are now equal also. Therefore, when the widths are equal, Prof. Gillespie's result, although obtained by much greater labor, is no better than the approx. result, while the content may still vary by $\tfrac{amD}{12}$ cu. ft. for every difference of m feet in the position of the fading end. Thus, the nearer the widths approach equality, the nearer is Prof. Gillespie's result to the mere approx. result, while the true content depends upon the place where the ridge or hollow terminates, not where it may be supposed to end. Taking the example of a right vol., mentioned in this connection by Prof. Gillespie, which is drawn to scale in diagram, and disregarding for the present the lines BG, GC, GI, the value of a is easily found to be 5.1. Substituting this in the formula $\tfrac{mD}{12}$, and making m 1 ft., D being 100, we find for the increment to the content $\tfrac{5.1 \times 100}{12}=42.5$ cu. ft., corresponding to an increment of 1 ft.

6

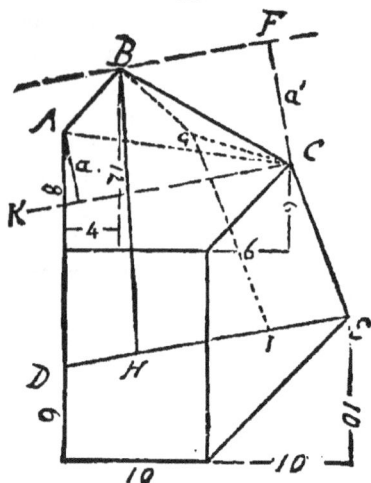

to d', measured along DE, or 43.4 cu ft. corresponding
to 1 ft. measured horizontally. Therefore, if BH happen
to divide AC, DE proportionally, Prof. Gillespie's result
is correct; but, if not, he errs by 43.4 cu. ft. for every
foot of distance from the true vanishing point to the assumed one. This is surely a difference worthy of being
noted, especially since it must be as easy to note the
true vanishing points in the field as to calculate the false
points by the rule of three. It was simply apparent to
Prof. Gillespie that some point must be taken for the
end H in order to obtain data for mid-section, necessary
to the prismoidal formula; but it seems that he was not
aware of the importance of finding the true position of
this end. This, however, might easily escape observa-
tion since it is a peculiarity of vols. bounded by warped
surfaces.

Prof. Gillespie's theory of the surface of all irreg. vols.,
whether bet. two irreg. cross-sections or one irreg. and
one reg., is summed up in the following sentence. "Conceive a series of vertical planes to pass through all the
points on each cross-section, at which the transverse
slope of the ground changes, and at which, therefore,
levels have been taken, and to cut the other cross-section
so as to divide the widths of the two *proportionally*."
By this treatment the surface of Prof. Gillespie's vol.
is scored by a number of ridges and hollows equal to the.

whole number of breaks in both end sections, against the great probability that the breaks at one end belong to the same ridges and hollows shown at the other : all the ridges and hollows of each cross-section vanish at the other, contrary in the highest degree to the natural disposition of the ground surface : a break representing a ridge at one end may be joined by proportional measurement with a break representing a hollow at the other, a perfect absurdity. His lines, fixed by an utterly unfounded assumption, are imaginary, as he calls Henck's diagonals, and lead to error nearly as gross. Since he assumes that each ridge or hollow vanishes in the same vol., he must calculate for each vol. the heights of as many vanishing ends as there are breaks in both end-sections, instead of calculating the heights of such only as do really vanish. After his laborious computations, first, to find the proportional distances, second, to find the height of one end of each ridge or hollow, third, from these to calculate the measurements of mid-section, the number of whose breaks is the number of breaks in both end-sections, generally double the number in the true mid-section, fourth, to obtain the areas of end and mid-sections, and, fifth, to apply the prismoidal formula, the result is identical with that obtained by approximating end areas, when the widths are equal, while the true vol. may vary far from this ; and the nearer the widths are to equality the nearer is Prof. Gillespie's result to the result of mere approximation. The ordinary methods of calculating irreg. vols., by approximating with end areas, and by finding heights of level cross-sections of equal areas and using tables, are both erroneous ; and assumption of any kind is entirely out of place for measurements which bear so important a relation to the content.

The formula $\frac{amD}{12}$ may be conveniently used as a correcting formula in the following case. Suppose a vanishing end has been neglected in the field, and it is afterward discovered by the relative dip of the lines AC, DE that this is an important point. Any point may now be assumed to satisfy the formula, for instance, the point D, already known, whose height need not be calculated, and afterward at a convenient time the distance m may be measured and the quantity $\frac{amp}{12}$ cu. ft. or $\frac{amp}{324}$ cu. yds. added to the former content. a is easily calculated, or can be found graphically and measured with sufficient accuracy. To determine whether $\frac{amp}{12}$ is positive or negative, conceive a line BF through B parallel to DE; then, if moving B to the right on this line increases the area of ABC, by lengthening the perpendicular, motion of H to the right also increases the content of the wedge. So, if D were assumed in order to avoid delaying calculation, and H were found to be the true point, $\frac{amp}{12}$ is positive. For a second break G, forming the second wedge $BCGEIH$, the formula of correction is $\frac{a'mp}{12}$, a' being FC or the distance of B from CK. If I were not known, it might be assumed at any point on DE, as E, H, or even D beyond H, and the correction made by measuring the distance m from the assumed to the true point. Since motion of G to the right on a line parallel to DE increases the end area and consequently the content of the railroad vol., so motion of I to the right increases the content. Whatever the position of breaks, motion of any in the same direction affects the content in the same way. If both H and I were for the present unknown, D might be assumed for both: then the cor. would be $\frac{ap}{12} \times \overline{DH} + \frac{a'p}{12} \times \overline{DI}$, DH being the increment m for the first and DI for the second. If D and E were assumed, the cor. would be $\frac{ap}{12} \times \overline{DH} + \frac{a'p}{12} \times -\overline{IE}$. To make these corrections m should properly be measured along the edge DE; but in the regular staking out of work the vanishing ends must, like real breaks, be fixed by horizontal measurement.

TABLES OF QUANTITIES.

As for *Tables of Quantities*, it is just to say that they are little used on many accounts. The common ones, for vols. bet. level sections, are reliable and expeditious; but these are the very vols. most easily calculated in series. For other sections, the areas of these must be independently calculated, and the centre-heights of equal level-sections computed or approximately taken from diagrams, as Trantwine's, before the tables can be used. This preparatory work is much greater than the labor of finding the whole content exactly by the method of this paper; and the level-section tables for all such cases are inaccurate, as illustrated by many authors and systematically discussed by Prof. Gillespie, the errors being almost constantly in defect and, therefore, seldom balancing. This error exists in all the tables constructed by followers of the methods of Telford and Sir John Macneill.

The bulk of a complete set of tables for all slopes and bases would be enormous, the number of such tables being equal to the product of the number of different bases by the number of different slopes in use. This bulk is sometimes reduced one-hundred fold by omitting the tenths, these to be interpolated by proportion, or by diagrams wherein the tenths are estimated by the eye. Such interpolation is not accurate. The diagrams are tiresome to the eye, and without the nicest care the producing values can not be laid off in the diagrams to the tenth part of a foot. An error of 0.1 in c, when $b=10$, $S=1$, $c=5$, and $D=100$, produces an error in the content of vol. $=0.1 \times 30 \times 100 = 300$ cu. ft., in case an equal error were made in the other end section. If no error were made at the other end, the error in content would be 150 cu. ft. If the error at the other end were equally far in the opposite direction, no error would exist in the result for the content. So, although there is a great chance that the errors may balance, it is not safe to use a method whose errors are so large. The same reasoning may be applied to the method of finding height of equivalent level-section. The measurements of irreg. section are in feet and tenths: it would rarely happen that the height of equivalent level-section would be exactly in feet and tenths. The neglect of the extra fraction, if only a fourth of a tenth, would make a difference in above example of 75 cu. ft., 37.5 cu. ft. or nothing in its several cases.

Some calculators have constructed auxiliary formulæ, by use of which vols. of certain base and slope can be treated by tables for other slopes and bases. These formulæ represent extra work. They may increase or diminish the error of the table, according to the respective slopes and bases.

The advantage of using tables of quantities is very much lessened by combining vols. in series, because thus all the constant factors are removed from the calculation of each. Suppose, for the sake of argument, we could have a table by which the approx. content,

$$\left(wc + w'c' + Sb(w+w') - 4Sb^2\right)\tfrac{D}{108},$$

of a vol., for a certain value of S and of b, could be obtained by mere reference, using some two measurements, as in the tables for level-sections. To find the contents of 20 such vols. consecutive, the table must be consulted 20 times and the results added. But, since the factor $\tfrac{D}{108}$ need be used once only for the 20 vols., the advantage of including it in the table is very slight. Therefore, if the table were constructed on the value within the parenthesis, the extra labor would be simply one little division. But the term $-Sb^2$ need be but once used. Therefore, if the table were constructed on the value $wc + w'c' + Sb(w+w')$, the extra labor would be only one subtraction and one division. Since the prod. $w'c'$ enters the formula for second vol., and every other mid-prod. is likewise common to two vols., and since the prod. $Sb(w+w')$ becomes for the series a single prod. of Sb by sum of widths, the formula $wc + w'c' + Sb(w+w')$ represents for the series 22 multiplications and the summing of 2 columns, the *prods.* and *widths*. Therefore, if we had a MULTIPLICATION TABLE extended enough for the prods. *wc* etc., to solve the above example we must consult the multiplication table 22 times, sum 2 columns, make one subtraction and one division. To accomplish the same by a table for the whole amount of each vol., we must consult the table 20 times and sum one column. The disadvantage, then, of using a table of prods. instead of using a table of contents for n vols. is composed of the following only.

The table of prods. must be consulted $n+2$ times instead of n times, an excess of 2. 2 columns must be summed instead of 1, an excess of 1. One subtraction and one division must be made.

These four extras represent very little work. The advantages of using a table of prods. instead of a table of contents are the following.

1. *Irreg. vols. can be treated by a table of prods. as well as reg. vols., because the formula for irreg. vols. consists likewise of a column of widths multiplied by Sb and a column of prods.*

2. *Since the irreg. and minor vols. can be included, n can be made greater, when the disadvantages of using the table of prods. becomes comparatively less.*

3. *Since the values S and b are not used in the construction of the table, it applies equally well to vols. of all slopes and bases, avoiding the great bulkiness of a complete set of tables.*

The above scheme of comparative merits of a table of prods. and a table of quantities is drawn up on the supposition that a table of quantities can be constructed for such vols. to be used directly. Since this cannot be done, as is evident by inspection of the formula which contains four variables, to the advantages of a table of prods. may be added, that it obviates not only bulkiness but the rules of equating areas, the diagrams, auxiliary rules and general inaccuracy. A table of areas might be constructed on the formula

$$\tfrac{1}{2}w(c + Sb) - Sb^2,$$

since it contains only two variables. This table must be consulted $n+1$ times, so the comparative disadvantage of the table of prods. would be a little less than named above, and the advantages would remain the same, since the table of areas must be reconstructed for every change of base or slope, etc. But the auxiliary rules, rules of equivalency, etc., would be avoided by either method. To correct the approx. results a table might be constructed on the formula

$$(w - w')(c' - c)\tfrac{D}{324};$$

but the table of prods. would still possess all the advantages, while the disadvantage would merely be the use of the factor $\tfrac{D}{324}$ once. The pris. cor. for irreg. vols., as well as the other, can be accomplished by a table of prods., and the factor $\tfrac{D}{324}$ used once for all.

We, therefore, strongly recommend the introduction and use of a table of prods. in connection with the calculation of railroad vols. in series. The multiplication table is seldom, if ever, used by calculators, for the reason that their work is desultory, changing every few seconds from one process to the other of addition, subtraction, multiplication and division; but in the method of computation recommended here, after transferring from Field-Book the widths and centres, we have in direct and uninterrupted succession a long line of multiplications to perform where the value of a table of prods. in easing the only laborious portion of the work will readily be recognized.

Such a table can be accommodated on 15 pages 8vo., ranging along tops of pages from 1 to 300, and down the pages from 1 to 80, thus containing all the ordinary prods. occurring in earthwork, of factors with three digits by factors with two, the decimal point to be inserted as required. For more digits than three the table can be as easily used. Suppose $w = 147.2, c = 26.3$. Look at top for 263; then down column to number opposite $14, = 3682$; then down further in same column to number opposite $72, = 18936$. Set in column of prods. $^{3682}_{18936}$. This is much shorter work than to transfer 147.2 and 26.3 to a separate piece of paper, find three partial prods., add these, and bring sum back to column of prods. By use of table no addition need be made till the column has been filled. All prods. made by factors near alike can be sought out at the same time. The table would be of use in any class of calculations, where the multiplications can be congregated.

FINIS.

FORMULÆ

FOR THE CALCULATION OF

RAILROAD EXCAVATION AND EMBANKMENT,

By J. WOODBRIDGE DAVIS, C. E.

Price (postage prepaid), in Stout Paper Binding, $1.00.
"   "   "   " Fine Cloth   "   1.50.
Colleges and Dealers supplied at the regular reduction.
To be had of G. S. ROBERTS. E. M., School of Mines, Columbia College,
N. Y. City, and of the principal publishers in this city.

BLANK SHEETS,

Ruled in columns, upon the plan of the TABLES OF CALCULATIONS in
this book, have been prepared for the use of calculators, who may adopt
this method. These sheets are 13x17 ins., of stout, heavy paper, like
cross-section sheets, and are neatly ruled in columns of proper widths
for the values they are meant to contain. The headings of the columns
are handsomely printed, as also the title of the sheets, viz.: TABLE OF
EARTH-WORK CALCULATIONS. These sheets are intended to be kept
on file or in port folio; and they entirely obviate the use of expensive
cross-section sheets, together with the labor of plotting, since the notes
of Field Book only are used.

One pattern, like that used on pages 47, 60, 63 and 75, for work containing intermediate stations, or irregular cross-sections, or both, in
fact, for any possible example, contains 40 lines, and serves, according
to the number of intermediates, for 10 to 30 stations, or 1,000 to 3,000
ft. The other pattern, like that on page 9, with the addition of the
column of prismoidal corrections, for work without irregularity or intermediate stations, on same size paper is ruled and printed double—
two similar sets of columns side by side. It, therefore, contains 80
lines, and would serve for a single cutting nearly 1½ miles long. Several
cuttings or fills can occupy the same sheet ; and, should an example be
too large for a sheet, the columns may be summed and the amounts
brought forward to a new sheet, and the example continued. ¾ in.
margin is left all around to preserve the work from rough handling in
looking through files.

Price (postage prepaid), For Interm. Stas., $5.00 per 100.
"   "   " without "   "   5.00 per 100.
Fractional parts of a hundred, down to ⅛, at same rate.
Amounts less than 20, 6 cents per sheet.
Assorted in any proportion of the two patterns.
Furnished only by G. S. ROBERTS, E. M., School of Mines, Columbia
College, N. Y. City, and by a few R. R. Stationers, who will send out
special circulars.

SCIENTIFIC BOOKS

PUBLISHED BY

D. VAN NOSTRAND,

23 Murray and 27 Warren Streets, New York.

WEISBACH'S MECHANICS OF ENGI-neering. Theoretical Mechanics. Translated from the fourth augmented and improved German edition, by Eckley B. Coxe. With 906 wood-cut illustrations, 8vo., 1100 pages, cloth, $10.

McCULLOCH'S ELEMENTARY TREA-tise on Heat. On the Mechanical Theory of Heat and its Application to Air and Steam Engines. 8vo., cloth, $3.50.

STONEY ON STRAINS. THE THEORY of Strains in Girders and similar Structures. 8vo., cloth, $12.50.

MacCORD'S SLIDE-VALVE. A PRAC-tical Treatise of the Action of the Eccentric upon the Slide-Valve. By Prof. W. C. MacCord of the Stevens Institute. 4to., cloth, illustrated, $3.

AUCHINCLOSS APPLICATION OF THE Slide-Valve and Link Motion to Stationary Portable, Locomotive and Marine Engines. 21 plates, 37 cuts, 6th edition, 8vo., cloth, $3.00.

IRON TRUSS-BRIDGES FOR RAIL-roads. The Method of Calculating the Strains in Trusses, with comparisons of the most prominent ones. By Col. Wm. E. Merrill. Second edition. 4to., cloth, $5.

SHREVE'S TREATISE ON THE Strength of Bridges and Roofs, with practical applications and examples, for the use of Students and Engineers. Illustrated with 87 wood-cut illustrations, 8vo., cloth, $5.00.

KANSAS CITY BRIDGE, WITH AN AC-count of the Regimen of the Missouri River, and a description of the methods used for Founding in that River. Illustrated with five lithographic views and 12 plates of plans, 8vo., cloth, $6.

CLARKE'S DESCRIPTION OF THE IRON Railway Bridge across the Mississippi River at Quincy, Ill. 21 lithographed plates, 8vo., cloth, $7.50.

WHIPPLE'S ELEMENTARY AND PRAC-tical Treatise on Bridge Building. 8vo., cloth, $4.

DUBOIS' NEW METHOD OF GRAPH-ical Statics. 60 illustrations, 8vo., cloth, $2.

GREENE'S GRAPHICAL METHOD FOR the Analysis of Bridge Trusses. Illustrated, 8vo., cloth, $2.

BOW'S TREATISE ON BRACING WITH its application to Bridges and other Structures of Wood or Iron. 156 illustrations. 8vo., cloth, $1.50.

HENRICI'S SKELETON STRUCTURES, especially in their application to the building of Steel and Iron Bridges. 8vo., cloth, $1.50.

STUART'S HOW TO BECOME A SUC-cessful Engineer. 18mo. boards, 50c.

HOWARD'S EARTHWORK MENSURA-tion on the Prismoidal Formula. Illustrated, 8vo., cloth, $1.50.

MORRIS' EASY RULES FOR THE MEA-surement of Earthwork. 78 Illustrations, 8vo., cloth, $1.50.

CLEVENGER'S TREATISE ON THE Method of Government Surveying. Illustrated, pocket form, morocco gilt, $2.50.

HEWSON'S PRINCIPLES AND PRAC-tice of Embanking Lands from River Floods. 8vo., cloth, $2.

GILLMORE (GEN'L Q. A.) ON THE CON-struction of Roads, Streets and Pavements. 70 illustrations, 12mo., cloth, $2.

GILLMORE'S (GEN'L Q. A.) TREATISE on Limes, Hydraulic Cements and Mortars. 5th edition, 8vo., cloth, $4.

BARBA ON THE USE OF STEEL IN CON-struction, Methods of Working, Applying and Testing Plates and Bars. Illustrated, cloth, $1.50.

SIMMS' PRINCIPLES AND PRACTICE of Leveling, showing its application to Railway Engineering and the Construction of Roads, etc. Illustrated, 8vo., cloth, $2.50.

HAMILTON'S USEFUL INFORMATION for Railroad Men. Pocket form, morocco gilt, $2.

SCHUMANN'S FORMULAS AND TABLES for Architects and Engineers. 307 illustrations, pocket-book style, morocco gilt edges, $2.50.

*** My Catalogue of AMERICAN AND FOREIGN SCIENTIFIC BOOKS, 128 pp., 8vo., sent on receipt of 10 cents.

www.ingramcontent.com/pod-product-compliance
Lightning Source LLC
Chambersburg PA
CBHW021413090426
42742CB00009B/1133